A History of Mechanical Engineering

A HISTORY OF
MECHANICAL
ENGINEERING

AUBREY F. BURSTALL

M.Sc. (Birmingham), Ph.D. (Cantab), D.Sc. (Melbourne)
Hon. D.Sc. (National University of Ireland)
M.I.Mech.E., Chartered Mechanical Engineer
Professor of Mechanical Engineering and
Director of the Stephenson Engineering Laboratories,
The University of Newcastle upon Tyne

FABER AND FABER
London

First published in 1963
by Faber and Faber Limited
24 Russell Square, London WC1
Reprinted 1970
Printed in Great Britain by
Latimer Trend & Co Ltd Whitstable
All rights reserved
©*Aubrey F. Burstall*, 1963

SBN (paper edition) 571 04707 6
SBN (cloth edition) 571 04706 8

The great thing about the past
is that it has happened.

A. J. P. TAYLOR in *The Observer*
2 August 1959

Preface

This book has been written for students of mechanical engineering. It is a review of events that the author considers to have been the most important in bringing the science and profession of mechanical engineering to its present state. Technical achievements which began in prehistoric times and continued throughout the ages at an ever-accelerating pace have been herein recorded and discussed. It can be looked upon as the minimum that every mechanical engineer should know about the past.

It is impossible in a book of this size to include all the multifarious advances that have been made in mechanical engineering, and the author's task has been to choose from the wealth of available material all that seemed most significant and appropriate for the young mechanical engineer, and to present it to him in a consecutive form with diagrams that would enable him to appreciate it. On this basis it appeared more important to dwell on, for example, developments in water turbines or lubrication rather than on textile or agricultural machinery, for the latter, though of daily impact and revolutionary importance when they took place, can now be seen to be out of the main stream that contributed to the knowledge that is the accepted heritage of the educated mechanical engineer. For this reason many important developments in other branches of engineering, viz. marine, aeronautical, automobile, mining and metallurgical engineering, have been omitted. Moreover, weapons of warfare and their development have not been included except where they contributed substantially to the development of mechanical engineering, though in a general sense the needs of warfare have always been a stimulus that released resources for experiment and invention.

There are nine chapters, covering successive periods of history. These periods become shorter as we approach the present day and the pace of advance accelerates. Each chapter includes sections on

(1) materials available to the mechanical engineer, (2) tools, (3) mechanisms and machines for the mechanical transmission of power and motion, (4) fluid machinery, (5) heat engines; these sections are preceded by some general comments on the state of science and civilization at each period and are followed by a general review of the progress and outstanding achievements of mechanical engineering at that date.

The author has not tried to assign credit to individuals for the inventions and achievements herein described. On the other hand, it has not been practical to avoid mentioning names altogether, many being so well known that they are almost part of the language of the engineer, and in this respect helpful both in definition and in recalling the important events of the past.

Great difficulty was experienced in trying to bring the assessment of events up to the present time in the last chapter. What seemed important even twenty years ago is not necessarily so today. In many instances recent innovations have no place here because there is doubt as to their value when a long term view is taken.

This book does not set out to record all mechanical inventions but only those which seem today to have been important in bringing the practice of mechanical engineering to its present state, nor have the social or economic consequences been analysed. Most of the recent mechanical innovations described have been commercially successful, since otherwise they would have remained technical curiosities and as such, though part of the history of invention, would not have been representative of the way in which mechanical engineering was carried on in practice. Some failures have been recorded because they are of technical interest, or for some special reason, but for the most part they have been omitted Another guide to selection has been that when some new mechanical principle was used successfully in a machine, it has been deemed worthy of inclusion.

The interest of the author in the history of his subject was first aroused in 1919, when as a student he took indicator diagrams on Watt's beam engine at Ocker Hill, Birmingham, during the James Watt Centenary celebrations, and thought he glimpsed the difficulties and triumphs of the pioneers. Later the works of I. B. Hart, H. W. Dickinson, C. Matschoss, R. J. Forbes, A. P. Usher and Lewis Mumford added fuel to the flames.

Thanks are due to the Council of the Institution of Mechanical Engineers for the help so generously given, and particularly to the Institution's present and former librarians. The *Proceedings* of the

Institution over the last hundred years proved a mine of information. For earlier periods, the five volumes of the monumental *History of Technology*, edited by C. Singer, were invaluable and to them the author gratefully acknowledges his debt.

While teaching the history of engineering to university students it has been found that lantern slides and cinema films are no substitute for models that 'work'. Models of such simple machines and mechanisms formed the basis of some of the diagrams in this book, and for this thanks are due to members of the laboratory and workshop staff of the Mechanical Engineering Department at King's College, Newcastle upon Tyne, for their skill and active co-operation.

Many of the diagrams were specially redrawn to the author's instructions by the departmental draughtsman, Mr. P. Elliott.

Finally the author would like to express his thanks to Dr. P. F. R. Venables, the general editor of this series, for his encouragement and for many useful suggestions, and to Mr. M. Shaw and other members of the firm of Messrs. Faber and Faber who have taken so much care in preparing the work for publication.

Acknowledgments

Acknowledgments are due to the following for permission to reproduce copyright material for the figures mentioned (the exact source is given in the List of Figures):

The American Machinist: Figs. 132, 135; *Antiquity*: Fig. 31; Edward Arnold (Publishers) Ltd.: Figs. 163, 164, 269, 272; Borg and Beck Ltd.: Fig. 207; B. T. Batsford Ltd.: Fig. 21; Chapman and Hall Ltd.: Figs. 79, 155, 218; *The Chartered Mechanical Engineer* (*C.M.E.*): Figs. 271, 287; Cincinnati Milling and Grinding Machines Inc.: Fig. 252; City of Liverpool Public Museum: Fig. 44; The Clarendon Press, Oxford: Fig. 36; Constable and Co. Ltd.: Figs. 158, 159, 161, 162, 165, 166, 212, 220, 222; Dobbie and McInnes, Glasgow: Fig. 274; Dr. A. G. Drachmann: Figs. 33, 34; Econ-Verlag, Düsseldorf: Figs. 104, 168, 286; *Engineering*: Fig. 233; T. Finkelstein and *The Engineer*: Figs. 184, 241; Goose and Son Ltd.: Fig. 235; Charles Griffin and Co. Ltd.: Figs. 223, 228, 275; George G. Harrap and Co. Ltd.: Figs. 180, 181; Dr. I. B. Hart and Methuen and Co. Ltd.: Figs. 45, 54, 64, 125, 169, 170, 188; Harvard University Press: Figs. 29, 43, 47, 49, 50, 52, 67, 71, 78, 87, 112, 142, 145, 146, 147; Herbert Hoover and Stanford University Press: Figs. 90, 107, 108, 109, 119; The Johns Hopkins Press: Figs. 30, 33; Institution of Mechanical Engineers (I.M.E.): Fig. 191; Iowa Institute of Hydraulic Research (Iowa Inst. Hyd. Res.): Figs. 18, 105, 106, 210, 213, 214, 264; Lawrence and Wishart Ltd.: Fig. 14; Longmans Green and Co. Ltd.: Figs. 203, 236, 237, 238, 273, 274, 284, 288; Longmans Green and Co. Ltd. for The British Council: Fig. 209; Macmillan and Co. Ltd.: Figs. 249, 250; The Massachusetts Institute of Technology Press (M.I.T.P.): Figs. 85, 97, 136, 137, 138, 140, 141, 193, 196, 197, 198, 199, 252; McGraw Hill Book Co. Inc.: Figs. 62,

82, 128, 154, 186, 195, 256, 270; The Executors of the late A. G. M. Michell: Fig. 261; Professor L. A. Moritz and the Clarendon Press, Oxford: Fig. 32 and Fig. 35 (based on *Saalburg Jahrbuch*, iii, p. 91, Fig. 45); Dr. J. Needham: Figs. 19, 20; Thomas Nelson and Sons Ltd.: Fig. 230; Max Parrish and Co. Ltd.: Fig. 183; Penguin Books Ltd.: Figs. 73, 95; *Philosophical Transactions of the Royal Society* (*Phil. Trans.*), London: Fig. 156; Sir Isaac Pitman and Sons Ltd.: Figs. 247, 248, 260; Prehistoric Society, Cambridge (Prehist. Soc. Camb.): Fig. 5; *Proceedings of the Institution of Mechanical Engineers* (*Proc.I.M.E.*): Figs. 38, 201, 202, 204, 211, 216, 217, 224, 229, 246, 253, 262, 266, 276, 277, 279, 280, 283, 290; Routledge and Kegan Paul Ltd.: Fig. 257; Science Museum (Sc. Mus.), London: Crown Copyright: Figs. 2, 124, 129, 133, 134; Science Museum (Sc. Mus.), London: Lent by University College, London: Fig. 8; Science Museum (Sc. Mus.), London: Lent by The English Electric Co. Ltd.: Fig. 234; *The Scientific American*: Fig. 187: The Smithsonian Institution; *United States National Museum Bulletin* (*U.S. Nat. Mus. Bull.*): Figs. 1, 26; Thames and Hudson Ltd.: Fig. 9; *Transactions of the Newcomen Society* (*T.N.S.*): Figs. 83, 84, 94, 98, 99, 101, 114, 115, 116, 117, 122, 167, 258; University Press, Cambridge (C.U.P.): Figs. 56, 121, 173, 174, 178, 185, 225, 231, 232, 239, 242, 243, 244, 278; University Press, Oxford (O.U.P.), Children's Department: Fig. 177; University Press, Oxford (O.U.P.) for Imperial Chemical Industries Ltd.: Figs. 3, 4, 6, 7, 10, 11, 12, 13, 15, 17, 22, 26, 27, 39, 40, 41, 59, 60, 63, 65, 68, 70, 75, 76, 77, 81, 88, 89, 92, 118, 127, 143, 144, 189, 190, 192, 226, 245; The Williams and Wilkins Co., Baltimore: Figs. 57, 58, 91, 100, 102, 110, 111, 113, 120.

Contents

Lists of references, and bibliographies, are included at the end of each chapter.

Illustrations

(for abbreviations see List of Acknowledgments)

Some of the figures, being among direct reproductions from the sources acknowledged, contain lettering which is not used in the text of this book.

The frontispiece is Fig. 126 in this list.

17

CHAPTER I

The Prehistoric Period, before 3000 B.C.

A bout 3,000 years before the birth of Christ, the Sumerians, a people living in the valleys of the rivers Euphrates and Tigris, in Mesopotamia, developed the art of writing, an invention that had great influence on the subsequent history of the world. There are obviously no documents or written records of what occurred before this and so the period up to 3000 B.C. is often termed prehistoric. What knowledge we have of man's achievements in prehistoric times has been acquired by examining tools and other buried objects that have survived on prehistoric sites in different parts of the world, and the sum total is now sufficiently large for archaeologists to be able to present to us a fairly clear picture of the mechanical equipment of antiquity though in many details it is still incomplete.

Many of the objects dug up by the archaeologist are preserved in museums where they can be seen and studied by successive generations of scholars and others interested in man's early history. Most finds relating to prehistoric (or 'Stone-Age') times consist mainly of tools and other articles made of stone. Wooden objects have seldom survived the passage of time and metal had only just begun to be used before 3000 B.C. Some articles of bone and pottery survive, and a few specimens of linen cloth and basketry.

Another source of our knowledge of prehistoric times comes from engravings on vases and ornaments and the remains of paintings found in tombs, caves and other sites that have been explored by archaeologists. From these we can in some cases infer the existence of mechanical contrivances for various purposes at very early dates. Some sites of buildings have ruins and foundations complete enough for us to reconstruct a picture of them at the time they were built.

The reader of ancient history will frequently come across the terms, the Stone Age, the Bronze Age, and the Iron Age. These terms refer to the materials that were in most common use for the making of tools and weapons by particular cultures of ancient man in different parts of the world at various times—thus no world-wide date can be given for the Stone Age, or the Bronze Age or the Iron Age. So far as the whole world is concerned, for example, a Bronze Age culture had been achieved in Babylonia by 3000 B.C. while Stone Age culture persisted for some time in other parts of the world—in Britain for another thousand years. A striking example of the length of time taken for a particular technique to travel the world is given by Childe who quotes the following dates for the first appearance of vases made on a potter's wheel: Sumeria, 3250 B.C., Palestine, 3000 B.C., Egypt, 2750 B.C., Crete, 2000 B.C., Greece, 1800 B.C., Italy, 750 B.C., Upper Danube and Rhine, 400 B.C., Southern England, 50 B.C., Scotland, A.D. 400, the Americas, A.D. 1550.[1]

MATERIALS

The conquest of fire was one of the earliest discoveries made by man in his search to understand the properties of the materials around him.[2] Once he had learned to make fire whenever he wished by rubbing or drilling or by percussion he was able to protect himself from the wild beasts and from the cold, and by cooking to extend his diet and preserve his food. Fire must have inspired primitive man with awe and wonder so that it is not surprising that fire-worship was often practised. One can imagine him marvelling at the hiss when water was dropped into the fire and at the breaking up of heated stones when dropped into water. The heating of different kinds of stones on the fire must have provided endless opportunities for experimenting, which no doubt was the beginning of extractive metallurgy. Likewise, experiments and observations of the effect of fire on clay hearths led the way to the firing of pottery.

Stone and wood were the materials, occurring naturally, that were most commonly used in engineering work. These were supplemented by materials taken from the bodies of animals such as bone, ivory, skin and gut. The greatest number of prehistoric objects that have survived are those of stone, and by studying them a great deal has been learned of man's first steps as a toolmaker, for it was stone, and particularly flint, that provided him with the hardest implements for cutting and shaping other materials such as wood and bone.[3]

By hammering stones on one another in different ways, even the hardest flint could be chipped or flaked to produce primitive knives, scrapers and axes, and by rubbing with softer stones a degree of what we should now call grinding and polishing was achieved. Small flint saws with serrated teeth were made for cutting small pieces of wood or bone. As time went on improved techniques were evolved for working stone to make implements so that the approximate date of stone tools that have been discovered can be estimated from the visible evidence of how they were made and their relative excellence of manufacture. L. S. B. Leakey describes in detail how flint was worked in prehistoric times by the hammer-stone technique, the prepared core method, the blade-flake method and by secondary flaking and by pressure flaking.

For making mechanical devices such as sledges and carts, wood was the principal material of construction and indeed it remained so until about 200 years ago when it began to be replaced by iron. For this reason Mumford has suggested that the rational conquest of the environment by means of machines is fundamentally the work of the woodman—whom he describes as the primitive engineer.[4] The physical properties of wood make it particularly suitable as a construction material for the primitive engineer: it can be drilled, split, sawn or carved and even hollowed out by burning; it is strong in tension, compression and shear, and it is elastic and light in weight; suitable pieces can be bent with the aid of heat; it is found in widely distributed places on the earth, both on the hills and in the valleys and in early times was found in abundance near lakes and rivers where primitive man so often made his home.

One of the earliest forms of joint, the mortice and tenon, was made in wood and was in use in a primitive form before 3000 B.C.

While metals were not in general use in prehistoric times there was one metal, gold—soft, easy to work and untarnishable—which was being worked to make jewellery and ornaments, and so the craft of the earliest goldsmiths laid the foundations of many metal-working processes that were later to be used for other metals on a much larger scale. Gold was the first metal to be cast, hammered into thin sheets and bowls, drawn into wires and joined together by soldering. The process of soldering was probably as well known 5,000 years ago in Ur as it is today in Europe.[5]

Tools before 3000 b.c.

The variety of tools made from flint and stone by primitive man was considerable. The scraper, the knife, the hammer, the anvil, the mallet, the wedge, the adze, the spokeshave and the saw are to be found among the remains of many primitive civilizations in different parts of the world. The earliest agricultural implement that has been found is a bone sickle set with flints in bitumen excavated in Palestine and believed to belong to about 6000 B.C. To the mechanical engineer it is the piercing tools, and particularly the drill, which have a special

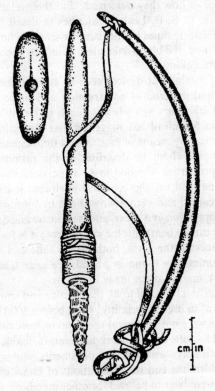

Fig. 1. Bow drill.

interest. It has been suggested that the early development of rotary motion by primitive man may have been connected with the ability of the hand, wrist and forearm to take part in twisting motions which would naturally occur if a scraper was being used to hollow out a piece of wood or stone. This must surely have been the earliest form of boring, though the motion involved was not continuous but partial rotary motion, as it was also with the bow drill and the pole lathe later. The bow drill in its various forms was common in many parts of the world, being used not only for drilling, boring and tre-panning but also for making fire.

The bow drill (Fig. 1) consists of four parts: a bent stave of wood with the bow string attached to its two ends and wrapped round the drill shaft in a simple loop in the centre, the drill shaft which carries a detachable bit forced into its lower end, and a socket in which it can turn at its upper end. The socket may be held in one hand—the other being free to work the bow—or in the mouth or even against the chest. A modification used by the Eskimos was the thong drill which required two operators, because in this version there is no bow, the bow string being replaced by a thong which is wound round the drill shaft, its two ends being held in each hand of one operator who pulls it back and forth while his assistant holds the drill in position by means of the socket.

A much simpler device than either of these sufficed for making fire in many parts of the world. A flat stick or hearth was held on the ground by the feet and a round stick was rotated between the palms of the hands while its lower end was pressed into a depression in the hearth—sufficient heat was produced at the point of contact to ignite the tinder placed there to start the fire.

In all these forms of drill, the motion was discontinuous, rotation being alternately in one direction and then in the other.

MACHINES

The bow and arrow has a special importance in the history of mechanics as it was the first contrivance made to store energy. Its origin is lost in antiquity. Forbes states that the first clear representa-tion of the bow is from N. Africa somewhere between thirty thou-sand B.C. and fifteen thousand B.C. when the present Sahara Desert was largely grass and parkland.

Many early bows were compound bows, that is, made up of two or three shorter lengths of elastic material which overlapped and were bound together with sinew.

Both the lever and the wedge were almost certainly in use in pre-historic times. One of the uses of the lever was in the early digging sticks used to cultivate the land. They were also used for moving heavy stones and other objects and in building construction and for excavating flints. The wedge was used for splitting wood and large rocks; in the latter case wooden wedges were often made to swell by wetting.

One of the most interesting machines devised by primitive man was the balance beam for weighing (Fig. 2). A prehistoric Egyptian

Fig. 2. Prehistoric Egyptian balance, 4500 B.C.

balance beam of red limestone $3\frac{1}{2}$ in. long with a set of stone weights, probably for weighing gold dust, has been found in a tomb at Naqada, Egypt. It is probably from the fifth millenium, 5000–4000 B.C. The beam was suspended by a cord in the centre and there were cords at each end, so there was no friction to impede the beam from swinging freely.

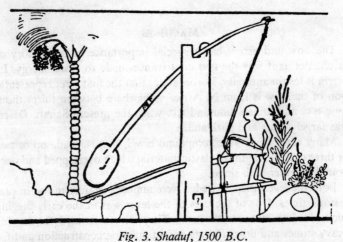

Fig. 3. Shaduf, 1500 B.C.

Another application of the principle of the lever existed in the 'swape' or 'shaduf', a long lever pivoted near one end (Fig. 3). A platform or water container was fixed at the short end from the fulcrum and balance weights at the long end, where a single man could raise several men or other heavy loads by applying his own weight to the end of the long arm. The shaduf was used extensively for irrigation and drainage and for lifting men over battlements during a siege. It was the most usual way of watering gardens in Egypt and Sumeria.

The plough was the earliest agricultural machine to be operated by animals. At first it was no more than a pointed stick—sometimes shod with flint—drawn through the ground by a rope tied to a pair

Fig. 4. Saddle quern, Egypt, 2500 B.C.

of oxen. Another agricultural device was the flail, for threshing corn. It has been surmised that this originated from the accidental breaking but not severing of a threshing stick which was found to be more effective with a loose hinge in the middle than without.

For grinding corn the saddle stone (Fig. 4) was used; i.e. one flat stone was rubbed back and forth on top of another, the corn being placed between them. The operator kneeled and used the whole body to push and pull the upper stone back and forth.

A device of even greater antiquity that is still in use today is the pestle and mortar. This could be made from a hollow or saucer-shaped stone for the mortar and a round stone held in the hand for the pestle or pounder used primarily for pounding or crushing grain, though the same combination could be used most effectively for mixing powder and liquid to form a paste. The ancient pestle and mortar is undoubtedly the ancestor of later crushing and grinding machines, particularly the rotary quern, the Greek trapetum and the later edge runner mill or mortar mixer.

Heavy objects were transported on rollers or sledges, and the oldest wheeled vehicle is a sledge on wheels known to have been in use in Sumeria soon after 3500 B.C. (Fig. 5). The wheels were solid but made of three wooden planks carved to fit segments of a circle and held together by transverse battens. It is not certain whether in the earliest wheeled carts the wheels turned with the axle or

*Fig. 5. Pictograph of wheeled vehicle
from Erech, Mesopotamia, 3500 B.C.*

rotated on it. Carts with two and four wheels are known to have been used and two-wheeled chariots were a particularly effective arm of Sumerian warfare both before and after 2500 B.C.

The history of the wheel is of particular importance to mechanical engineers since so many industrial applications of rotary motion have derived from it. Childe states that discs revolving freely on a fixed axis or with an axle free to turn in a bearing were in use both in the ceramic and transport industries between 3500 and 3000 B.C.

The potter's wheel (Fig. 6), a turntable moving about a vertical axis, has been used from that time onwards to lessen the labour of working clay into circular forms in making pottery, and to increase the speed of manufacture and the circularity of the product. By throwing clay accurately on the centre of a rotating turntable, centrifugal force can be used—with only a light pressure from the potter's hand—to make the clay assume any circular form that may be desired. The wheel is revolved by giving the edge a push at intervals with a free hand or with the foot.

Most of the ancient potters' wheels were made of wood and their existence has to be inferred from the surviving fragments of pottery that were obviously made on a potter's wheel. Some wheels were made of clay or stone, the bearings consisting of pivoted and socketed stones in pairs. One of the earliest was discovered by Woolley in his excavations at Ur.[5] One that he found in the 'Uruk' stratum at Ur (3250 B.C.) he describes as 'a heavy disc of baked clay about three feet in diameter with a central pivot hole and a small hole near the rim to take a handle'. A number of stone sockets for door pivots have also been found. It appears that in prehistoric times and in places such as Mesopotamia where stone was scarce, door sockets

Fig. 6. Potter's wheel.

were regarded as valuable articles—not as part of a house, but rather as furniture that a tenant was expected to bring with him. The doors were swung on poles projecting above and below the door frame, the lower end resting in the door socket embedded in the floor.

Although woven fabrics were made much earlier than 3000 B.C. no machinery had been developed for spinning, which was all done by hand. It was done by twirling a straight spindle held vertically between thumb and forefinger while the threads were drawn out and wound onto the spindle. The spindle was usually fitted with a collar of stone, clay, or bone, and this collar, known as the whorl, acted as a flywheel to maintain the spin. Many of these whorls have been found in excavations of primitive cultures; for example 8,000 were found in the earliest excavations of Troy and many more of later date in Cyprus, Palestine, Egypt and in the Lake village cultures in Switzerland.

The earliest representation of a loom is of a simple horizontal loom, on a pottery dish which has been dated 4400 B.C. from Badari, in Egypt (Fig. 7). In this type of loom the warp is stretched between two beams that are fastened to four pegs driven into the ground. It has a shed-rod, a rod heddle and a primitive shuttle, though a needle was sometimes used as a shuttle.

Traces of these looms and samples of textiles woven on them have been found on many ancient sites.

Rope was important to primitive man for fishing lines and the making of nets and snares; also by its aid loads could be dragged from place to place. An early cave painting in Spain shows a rope being used to let down a honey-gatherer from a cliff. In the ancient empires it was only by rope-pulling that manpower in large numbers could be employed, as in the transport of colossal statues or the

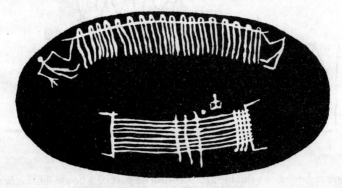

Fig. 7. Loom on dish, 4400 B.C.

building of temples. Ropes several inches thick, of three-strand papyrus, have been found on quarry floors. Each strand had 40 yarns of seven fibres.

Various raw materials were employed according to their availability. Early ropes of hair, sinew, twisted fibres, strips of hide or papyrus reeds, and of date palm fibre, flax or halfa grass, were made by twisting or plaiting. Rope was needed for ships' rigging, and for slings, bow-strings, lassos, bolas (bunches of weighted cords, thrown to entangle game), nets and lines. Strong fine flaxen rope from early times is found in Egypt. Glanville says rope of Eleventh Dynasty compares well with our modern products. Lashings for binding together furniture and for building graves were also in demand in Egypt.

In later developments rope was an essential component of early machines, such as the bow drill and pole lathe, for transmitting motion and energy, and it retained its importance for many centuries.

REVIEW

This review of the slender mechanical resources of prehistoric times indicates the limitations that were imposed by the primitive—though fundamental—character of the tools and implements constructed first of stone and later of bronze. Many of the simple hand tools that we use today, such as the hammer, the axe, the saw and the drill, were already in use in those early times. The principal mechanical achievement was tool-making and the most advanced tool was the bow drill.

Two outstanding discoveries in the field of engineering material were the ability to make fire and to melt and work metal—in particular, gold.

We can dimly perceive the beginning of machines in the wedge, the lever and the wheel, all of which were in common use before 3000 B.C. A particularly important application of the lever principle was the balance beam for weighing. Here was the beginning of measurement—and of the experimental method—for the results of the weighing could be repeated any number of times with certainty that the same result would be obtained.

Nothing that could properly be called a machine appears to have existed. No fluid machines or heat engines or any parts of these machines had yet been invented.

Though a large part of the time of these ancient peoples was taken up by food gathering, the daily tasks of primitive agriculture and the domestication of animals, in the great river civilizations of Egypt and Mesopotamia the population was so large and the soil so fertile that there was a surplus of food that could go to supply the craftsmen, priests and officials and all who were engaged in pursuits other than hunting or agriculture, and on whom the future progress of their culture depended.

REFERENCES

(for abbreviations see List of Acknowledgments)

1. Singer, C. (editor), *A History of Technology*, vol. i. O.U.P., 1954.
2. *The Bryant and May Museum of Firemaking Appliances* (and supplement). Sc. Mus., London, 1926, supplement, 1928.
3. Petrie, Sir W. M. Flinders, *Tools and Weapons*. University College, London, 1917.

4. Mumford, Lewis, *Technics and Civilization*. G. Routledge and Sons, London, 1934.

5. Woolley, Sir L., *Excavations at Ur*. Ernest Benn, London, 1954.

BIBLIOGRAPHY

Forbes, R. J., *Man the Maker*. Constable, London, 1950.

Hawkes, J. and C., *Prehistoric Britain*. Chatto and Windus, London, 1947.

Sarton, G., *A History of Science*. O.U.P., 1953.

Singer, C. (editor), *A History of Technology*, vol. i. O.U.P., 1954.

CHAPTER II

The Period of the Egyptian Empires, 3000 B.C.–600 B.C.

The history of mechanical engineering becomes less a matter of conjecture and takes on the character of historical truth with the invention of writing which occurred somewhere about 3000 B.C. This is attributed to the Sumerians, who, as the Egyptians had done in the course of developing their hieroglyphic writing, began with pictographs, from which they developed live characters and wedge-shaped signs which made up cuneiform script. These characters could be inscribed with suitable tools on hard rocky surfaces and more easily on clay or pottery before it was fired. The Sumerians, who had large supplies of clay in Mesopotamia, soon developed the use of clay tablets for writing, using a reed to make the marks in the soft fresh clay, which was then dried or baked to last indefinitely. Cuneiform script was peculiarly well adapted to writing on clay and it continued in the Near East wherever clay was obtainable for nearly three thousand years. Many thousands of cuneiform tablets have been found during excavations. For example, American excavators at Nippur, a Sumerian religious centre, found a large library of tablets and 'text-books' for teaching students at the time of Hammurabi (1728–1686 B.C.). Important documents have survived in exceptionally good condition because they were enclosed in baked clay envelopes which had to be broken to extract the tablets.

Other ancient peoples wrote on wooden tablets, parchment and leather, which are more perishable materials, though a few have survived because of the exceptional conditions in which they were stored. The Egyptians wrote mostly on papyrus, much of which has survived because of the exceptionally dry climate of that country. The sheets of papyrus for writing were made by cutting the pith of the reed into thin slices and then laying the slices together with glue and beating several layers upon one another with wooden mallets. Reeds of small diameter were used as pens, frayed at the end like a brush. Before writing this was dipped in water and rubbed on a cube of ink.

Sarton[1] mentions that about one hundred clay tablets are known, mostly obtained clandestinely, that refer to Sumerian mathematics and tables of numbers. These include tables of squares and cubes, square roots and cube roots, reciprocals and multiplication tables. The Sumerians had originated a decimal system. They seem to have

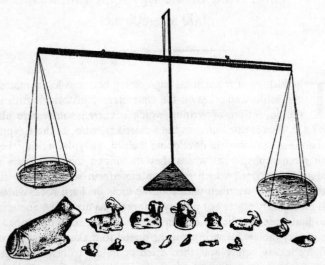

Fig. 8. Standard weights, El Amarna.

had a natural genius for algebra and were certainly able to solve linear, quadratic and cubic equations. Their most surprising achievement was their handling and understanding of negative numbers, a concept that did not penetrate Western minds until centuries later.

Standards of length and weight probably date from about 3000 B.C. when they were kept in the temples and carried the authority of the King.[2] The Ancient Egyptians standardized units of length derived from parts of the human body—the *great span* from thumb to little finger—*the foot* and the *cubit*—the last being the length of the forearm (20·6 in.). There were also palms and finger widths. A fathom was four cubits and cubit rods can still be seen in museums. The standard cubit length was maintained plus or minus 2 per cent from 3000 B.C. to the mid-nineteenth century A.D. and then almost wholly replaced by the metric system. It was used in Europe for centuries as a measure for cloth in the Middle Ages.

Standards of weight were used commercially in India, Mesopotamia and Egypt about 2500 B.C. (Fig. 8) but weights and balances were used by jewellers before this for weighing gold dust. Standard weights made of stone or bronze, in the forms of birds such as the sleeping duck of the Sumerians, and animals such as the crouching lion of the Assyrians, were used in conjunction with a balance. Small units were known as *shekels*, larger ones as minas and talents.

Standard units of capacity or of volume—the horn—were used for trading in oil, wine, gold and linen, but there was much local variation.

MATERIALS

During the period from 3000–600 B.C. the ancient Egyptians perfected the working of stone. Stone was used not only for tools—before its use for such purposes gave place successively to copper, bronze and iron—but it was used on a vast scale as a constructional material and for carving statues and vases. Their mastery of the medium is shown by a picture of a stone vase having its interior drilled out by hand with a trepanning tool. To hollow out a piece of stone to make a necked vase is obviously a very difficult and lengthy mechanical operation. The time can be much reduced by trepanning, that is, drilling with a hollow drill made in this case of a piece of copper tube, the abrading operation being assisted by sand and water and resulting in the removal of a solid core.

Soft rocks were often cut with a chisel and mallet or sawn, but hard rocks were pounded with balls of stone or hand hammers made of dolerite, and drilled and sawn, sometimes with the aid of an abrasive material such as quartz sand. Flinders Petrie instances, for accuracy of grinding, the granite sarcophagus of Senusert II of 3350 B.C. described as 'ground flat on the sides with a matt face like ground glass that only has about a 200th of an inch error of flatness and parallelism of the side'.

Traces of wedge slots in the granite in the Aswan quarries supply evidence of the use of wedges in ancient stone-winning. Copper and also wooden wedges were used, the latter being wetted so that some ten hours or so later, when the wood had expanded, the rock would split.

Very large blocks of granite were quarried in the mountains east of the Nile—to be used in the construction of the temples, pylons and tombs that were the striking architectural features of the Egyptian civilization and commemorated the glorious reigns of the

Pharaohs and the gods they served. The pyramids were tombs of kings of the third millennium B.C. Expeditions were sent to distant quarries, seeking stone, and one such expedition, in the time of Rameses the Great, involved the assembling of nine thousand men. Five thousand were soldiers, two thousand were temple staff and eight hundred were foreign auxiliaries. There were also nine hundred officials. They were required to convey safely one artist, three master quarrymen, a hundred and thirty quarrymen and stone dressers, two draughtsmen and four sculptors. Most of them were needed to transport the stone or carry supplies.

The implements used in stone building were the lever, ropes, rollers, sleepers, plumb lines, sledges and rafts and unlimited man-power. For levelling, water was used, poured into a clay trough fixed round the edge of the object to be levelled. Some authorities believe that for raising large stones vast embankments of rubble were built temporarily and removed before the work was finished. Traces of such ramps have been found near several large monuments. Another theory has been advanced concerning the erection of the large pyramids in which it is suggested that the larger blocks were raised from below by balancing them alternately on two fulcrums near the centre, the block being tilted by pulling with ropes or by large numbers of men standing on the block, moving from one end to the other, so that the fulcrum which was free could be wedged a little higher after each displacement. Yet another theory suggests that the sides of the pyramids were themselves used as ramps, the facing being completed last from the top downwards. The Egyptians did not use the pulley.

More than two million blocks of limestone were used in constructing the Great Pyramid in Egypt, the largest block being more than one thousand tons. It was recorded by Herodotus long afterwards that 100,000 men worked on this project for twenty years. Many were craftsmen and farmers, the latter working on the job in lieu of paying taxes during that part of the season when their land was under water during the Nile flooding. The quarrying, transport, erection and finishing of the material must have involved a great deal of organization in order that such large numbers of men could work effectively.

Many conjectures have been made as to how the large blocks of stone comprising the pyramids were finally slid off their sledges or supports into position. It seems most likely that they were floated on a thin bed of wet viscous mortar, using ropes, wedges and levers—surely one of the first useful applications of fluid friction as we understand it today! The ancient Egyptians also had some appreciation of the difference in the limiting load that stone could carry in bending and in compression and had apparently discovered that the limestone from Tura was unsuitable for free-standing columns which had to support a heavy roof, and that roof slabs of the same material should not span a width of more than about nine feet. Consequently the passages and chambers of some of the early Egyptian (Zoser) masonry are very narrow. It was evidently appreciated that there was some practical limit to the load that could be carried by the columns and beams used in building construction. Forbes states categorically that it is certain that structural drawings of the pyramids existed.

WOOD

With the advent of metal tools in the bronze age and later in the iron age the arts of wood-working became more precise and more specialized. Carpentry flourished. By 2800 B.C. it could be said that a complete mastery over wood had been achieved in the most advanced civilizations. Elaborately decorated boxes were being made, and halved and mitred joints were being used. Plywood had been invented in Egypt where wood was scarce and plywood with six alternate layers of different varieties of wood was made. The plies were usually fastened together by wooden pegs. Twelve centuries later, in the Eighteenth Dynasty, veneering was practised and inlaying with ivory and ebony and overlaying with gold, silver and copper had been perfected. Nails were usually wood pegs but for delicate work minute metal nails were used. The adze for smoothing the surface of wood was one of the principal tools of the carpenter, with a blade up to 18 in. wide, for the wood plane was not invented until Roman times.

The increasing skill of the worker in wood is particularly evident in the wooden carts and chariots that have been discovered. At this period over most of the known world it was the custom to bury the dead accompanied by their choicest possessions, for use in the next world. These, where the tombs were not robbed, remained unspoilt (owing to the dry climate in Egypt), until unearthed in modern times. In the Crimea, burials were preserved by the heap of stones covering the grave becoming capped with ice and snow, which sealed them off until they were excavated. These 'frozen graves' were made by the Scythians, an Indo-European transcaucasian nomadic tribe that migrated west and flourished from the seventh to the third century B.C. In Egypt chariots from the tombs had light spoked wheels and each chariot had more than fifty component parts. Among the Scythians, who had many horses and had mastered the technique of riding, carts with four-spoked wheels up to six feet in diameter, and with central poles for yoking two, four or eight horses, were used for transport and could be tented over for bivouac. The war chariot was lighter, faster, and more easily manœuvrable. Its spoked wheel, which came into use at the beginning of the bronze age, could not have been made without a metal saw (Fig. 9).

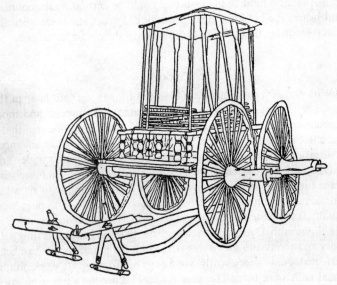

Fig. 9. Scythian cart.

The woods used in the ancient empires were mostly imported in historical times from W. Asia. Beech, box, cedar, cypress and fir, have all been identified from shrines, coffins, labels on mummies and other objects in graves. They also used juniper, lime, oak, pine and yew. Willow was used for handles and poles. There was lively appreciation of the virtues of each type of wood for the purpose in hand.

Other materials used for engineering purposes were leather—for ropes, fastenings, bellows, hunting slings, buckets, chair seats, chariot floors, tyres, harness, dagger sheaths; bone, horn and ivory for awls, rings, pins and needles and for handles for tools; glue and animal hair for making brushes; rawhide used for fastening handles to tools and to make bearings for bronze chariot axles. Emery and pumice were used for polishing, quartz sand as a common abrasive; gypsum plaster was used in Egypt in preference to lime mortar because fuel was scarce and lime-burning needed three times as much fuel as gypsum which was found to be adequate in the dry climate; bricks were made of clay and reeds and dried in the sun, until Roman times when kiln drying began.

In Mesopotamia, bitumen occurred naturally as seepages in valleys and as rock asphalt in the mountains. In very early times it was increasingly used for channels, waterproofing, and for the matrix in which mosaics of coloured stone or mother-of-pearl were set, to form a decoration for walls or floors. It was important in antiquity as a building material and as a paint for waterproofing porous bricks, and continued thus until Greek and Roman times, when building methods changed, and the smaller demand could be more conveniently met by wood-tar and pitch, so that ancient sources were forgotten. Bitumen was also used to plaster coracles, for lining domestic drainage systems, bathroom floors, etc., for caulking wooden boats and for the outer coating of grain bins. In Ptolemaic times it was used occasionally in the mummification process.

METALS

Gold. Gold was much sought after. New sources were discovered for the increasing quantities needed in the ancient empires of Egypt and Mesopotamia. Expeditions were sent long distances in search of it, finding it usually in quartz veins running through granite. The gold-bearing rock was pounded with dolerite balls to break it up, and the workings sometimes extended for hundreds of yards into the rock face, where traces are still visible.

Some of the gold found occurred naturally with silver, which gave it a lighter colour, but it could still be melted and worked by the goldsmith in the usual way. Very large quantities of gold were found and used in the ancient empires[3]; e.g. the weight of gold in the coffin of Tutankamen (1350 B.C.) found by Carter was 300 lb. In Egypt gold was more plentiful than silver, but in Mesopotamia the reverse was the case. Impure gold containing a large proportion of silver was called electrum.

In the working of malleable metal, it was the goldsmith who laid the foundation of modern metallurgy. He was the inventor of swageing and wire drawing, drawing his wire not only in round sections extremely fine, but also in square sections, which were then twisted to give a rich effect for ornament. Sequins were stamped out, as were coins, and sheet gold was beaten out to make all kinds of vessels, ornaments and gold leaf. Gold rings have been found in tombs from a very early date and in at least one case chains have been found of gold rings made presumably by cutting and hard soldering. The ingenious and ancient 'cire perdu' or lost wax process of casting was first practised by the goldsmith. In this method a wax model is made of the object to be cast; the model is then covered with clay and baked, whereupon the wax runs out of the mould leaving it exactly of the shape desired. The casting is then made by pouring the molten metal into the baked mould, which when cooled is broken away leaving the casting intact.

Copper. Small amounts of native copper were found and used but the development at this period of most importance to engineering was the smelting of metals from their ores.

Copper was being smelted in Mesopotamia before 3000 B.C. and was in common use in Egypt from 2600 B.C. It has been estimated that during the next 2000 years many thousands of tons of copper were produced in Egypt, and at the end of this time it was being imported from Cyprus, and from many parts of Europe. The ore was mined underground by fire-setting, that is, digging down a rock face and then lighting a large fire at the bottom of the hole so that the rock face was heated until it broke up, water often being poured onto the hot face at intervals to assist the disintegration by setting up temperature stresses. Smelting consisted of placing the ore with charcoal into a furnace and blowing air through it. At first the blowing was done with pipes in the mouth, and later by bellows worked with the hands or feet. Ladles of crude pottery were used to pour the molten metal into moulds of stone or dried clay. The ladle or crucible was held in the bent crook of a pliant sapling—a primitive form of spring tongs (Fig. 10).

The casting and working of copper soon gave place to the making of bronze—i.e. an alloy of copper and tin. This was easier to cast and much stronger in use. Both copper and bronze were commonly hardened by cold working, that is by repeated hammering of the metal on an anvil while it is cold.

The period during which copper and bronze were the materials used for making tools and weapons covered about two thousand years in Mesopotamia and Egypt.

Iron. The Iron Age began in the East in about 1200 B.C. and the discovery of iron is generally attributed to the Hittites, who lived in the mountainous region of what is now known as Asia Minor. For long before then odd pieces of iron had been prized as rarities but these were probably obtained from meteorites.

The smelting of iron from its ores was essentially a much more complicated and involved process than the smelting of gold, silver, copper or their alloys. In the first place it required much higher temperatures. Iron melts at 1,530° C., compared with 1,083° C. for copper, 1,060° C. for gold and 960° C. for silver. Accordingly a

Fig. 10. Casting bronze door, Egypt, 1500 B.C.

greater blast of air was required and the furnace in which this operation was done eventually became known as a blast furnace. Secondly the right flux was needed to ensure the proper slagging of the ore. Lime and sometimes limestone were used. The bloom or product from this furnace contained cinders, charcoal, and pieces of slag, so it had to be re-heated in another furnace and hammered many times while hot to remove this non-metallic matter and to consolidate the metal particles into a solid billet. Clearly these operations required that a number of tools be developed—tongs for handling the red-hot metal, hammers and anvils, crucibles, furnaces and bellows; all these items of equipment had reached an advanced stage of development by 1200 B.C.

The major metallurgical triumphs of the Iron Age blacksmiths were the three discoveries of (1) steel-making by carburizing, i.e. repeated heating of the iron billet in contact with charcoal; (2) the hardening of steel by sudden quenching from a high temperature; and (3) the tempering of hardened steel by slow heating to a moderate temperature to avoid brittleness. The last two discoveries were made successively much later than the first, hardening of steel about 800 B.C. by the Greeks and tempering of steel in Roman times. Cast iron was unknown in Ancient Egypt and in Greece and Rome, probably because the furnaces had never reached a sufficiently high temperature really to liquefy the metal.

TOOLS

The advent of metal revolutionized the making of hand tools, and some changes in design resulted. The discoveries first of bronze and later of iron brought immediate and profound improvements in

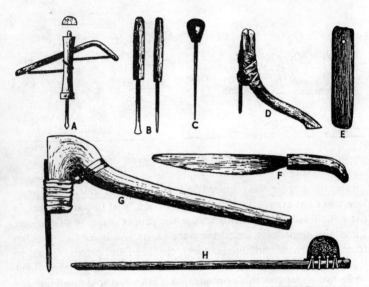

Fig. 11. Egyptian wood-working tools, 1200 B.C.

the quality of hand tools (Fig. 11). The problem of fixing the wooden handle was solved by leaving a hole in the metal head through which the handle could be passed, to be fixed by wedges in the manner that is used up to the present day. In fact, before this period was over hand tools for working in wood were in a great many respects the same as those we have now. Nearly every iron tool had some improvement in design as compared with the bronze tools; for example wood chisels were heavier and the wooden handles were provided with ferrules, and some of the iron saws had teeth which were raked to prevent jamming of the blade and to assist in the continuous removal of sawdust. A hoard of Assyrian tools found at Thebes in Egypt by Petrie and dated the eighth century B.C. includes an iron rasp for woodworking, the ancestor of the present steel file for metal. Drilling by hand was improved by the introduction of various improvements to the bow drill: an Egyptian wall painting shows three drills being operated by one man with a single bow while drilling beads (Fig. 12).

Fig. 12. Triple-drill Egyptian beadmaker, 1450 B.C.

Another improvement was the use of weights fixed to swinging arms for heavy drilling in stone, usually with a copper bit in the form of a tube. Some abrasive and perhaps water was used, and by trepanning in this fashion much of the labour of hollowing out a stone vase was reduced. Flint bits were also used for this purpose but soon after the start of the Iron Age in Egypt, arrow-headed iron or steel drills were invented and remained in use until they were superseded by twist drills in the present century.

The invention of the vise had to wait for that of the screw and at this period the carpenter held his work firm either by lashing it with thongs to his bench or to a vertical pillar, or it was secured by pegs and wedges driven into holes in his work bench (Fig. 13). No date

Fig. 13. Joiners at work.

can be given for the origin of the lathe. Examples of turning appear in many places in the Eastern Mediterranean from about 1500 B.C. and are common by 900 B.C. Lathes for turning wood are mentioned by Plato, 400 B.C. It seems certain that the lathe in its most primitive form—the bow lathe—developed from the bow drill, the object being turned replacing the stock of the drill. A later important development was the use of a cantilever spring to provide the return motion so that the operator had only to pull the bow string and could therefore dispense with the bow. For a cantilever spring a young sapling or pole firmly fixed by its roots or in some other way at one end was used. The other end was pulled down by the string as the operator rotated the work. This type of lathe—known as a pole lathe—is still in use in India and in parts of the East.

For handling materials, chains as well as ropes were used. Chains made from rings each folded in half and then passed through the next link were found in the Early Cemetery at Ur, 2500 B.C. Chains with S-shaped links are found from the eighth century B.C. at Nimrud where there were also heavy chains of cast bronze links.

Mechanisms

Two of the greatest mechanical achievements of this period were the inventions of the windlass and the pulley. The windlass is believed to have been first used by the Bronze Age miners in an attempt to lessen the labour of bringing ore up the mine shaft, but it may have originated where we can sometimes see it now—at the top of a well for drawing up the wooden or leather bucket, in those times a daily task.

There is every reason to believe that the Bronze Age Egyptians did not have the pulley, for it is never shown in the pictures of their vast building operations or for hoisting the sails on their ships which retained many primitive constructional features. The first pictorial representation of a pulley is in an Assyrian relief (Fig. 14) in the eighth century B.C. The pulley was a tremendous mechanical achievement for it enabled a downward pull to be used to create an upward force.

In the field of transport the two-wheeled chariot and the four-wheeled cart were both much improved by the spoked wheel, an immense improvement over the small disc-type wheels used before. Spoked wheels were lighter and could be made of much larger dia-meter—up to 6 ft. or more—so that the chariot or cart could be driven at much higher speeds over rough ground—a feature that won many battles when chariots with spoked wheels were first intro-duced. Figure 15 shows a complete chariot recovered from an Egyptian grave (1500 B.C.). Fatty matter was used to lubricate the axles of chariot wheels and its use for lubricating rawhide bearings

Fig. 14. Pulley, eighth century B.C.

Fig. 15. Thebes chariot, 1500 B.C.

continued until a hundred years ago. The development of harness for horses and oxen, to employ the power of their muscles in drawing vehicles and agricultural implements and in operating machinery, must have occurred at a very early date as can be seen from murals. However, the harness designed primarily for oxen did not allow the horse to use its strength to the best advantage. It should be able to lower its head to exert its full strength and the bearing rein prevented this.

The first rotary machine made its appearance during this period— the rotary hand mill or quern, for grinding corn, which was known in Syria before 1200 B.C. (Fig. 31).

FLUID MACHINES

The origins of fluid machines are to be found in this period as in pipes for conducting liquids and for blowing air; in the blow gun and the syringe which are probably the first examples of the piston and cylinder; in the bellows which played such an essential part in the smelting of metals; and in the discovery of the siphon, the theory of which was to remain unexplained until the middle ages.

Fig. 16. Water clock.

The Egyptians used the flow of water to measure the passage of time. Their water clocks (Fig. 16) show some understanding of the principles of water flow, for the stone vessels containing the water were made with sloping sides in the shape of a truncated cone, the water being allowed to escape slowly through a small hole at the bottom of the vessel, the time being shown by the level of the water inside the vessel, which was graduated with a uniform scale to show the hours. With this shape of vessel the water level falls almost uniformly because the greater flow when the vessel is full is compensated by the greater cross-sectional area at the top.

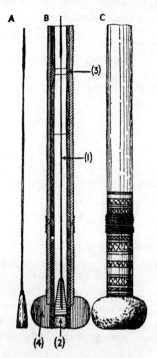

Fig. 17. Blow gun, A.D. 1900.
BC

The blow gun (Fig. 17) was used for killing at a distance with a wooden dart having a poisoned tip. The barrel was a tube of tropical reed or bamboo between 6 and 12 ft. long and about 1-in. internal diameter. The butt of the dart was secured in a plug of pith to prevent air leakage—probably the first form of piston packing.

Little is known about the piston-type syringe beyond its illustration in Egyptian records in connection with the art of embalming or mummification. Both the pipette and the siphon are shown in Egyptian pictures illustrating the transfer of liquids from one container to another (Fig. 18). There is also a copper drinking tube from Eshnunna, Mesopotamia (2300 B.C.) shown on a cylinder seal.[4]

There are many pictures of bellows in use for smelting and forging. The earliest are hand-operated but in Ancient Egypt foot bellows, one for each foot, made it possible to supply a steady continuous blast of air. Some illustrations (Fig. 10) show the bellows being pulled up on the inlet stroke by a thong held in the hand and attached to the upper surface of the bellows. However, it has been suggested

Fig. 18. Siphon tubes, 1450 B.C.

that these were later improved to provide for being driven and valved simultaneously by foot manipulation. (On the other hand Farrington believes that forge bellows with wooden top and bottom boards and leather flap valves were probably only invented in the twelfth century.)

The counterbalanced baler or shaduf was used for raising water for irrigation purposes throughout this period, but in the last few centuries before the classical period the chain of pots (or Persian wheel) was developed (see Fig. 45)—probably used for watering the hanging gardens of Babylon (700 B.C.). Whether this was operated by manpower, by animals or some other means, is obscure.

REVIEW

The most noteworthy development of this period was the appearance of crude machines to perform some of the more laborious tasks of mankind. There were the windlass and the pulley for raising water from the well and for lifting generally, the chain of pots for irrigation, the rotary quern for grinding corn and the bellows for smelting. All of these were operated by manpower, and most were based upon the wheel, which had been improved for chariots by becoming spoked. An essential feature of the rotary quern was its cranked handle and this was the first appearance of an essential machine element that was to be embodied in many later machines. Another machine element of equal importance was the piston working in a cylinder which was embodied in the syringe and the blow gun.

Probably the most sophisticated machine of the period was the bellows, since it involved the use of valves, operated either automatically by leather flaps, or by hand. In either case it depended on the periodical opening and closing of an air passage at the right time in a series of recurring events, and this conception represents a great step forward towards the construction of self-acting machinery.

In the copper tubing that has been found and in the ancient water clocks we see the beginnings of control of the flow of liquids to achieve a given purpose, an essential first step that had to be taken in the development of fluid machinery.

The mechanical engineer of today is perhaps most impressed by the legacies of Sumeria and ancient Egypt in regard to measurement—arithmetic, algebra and geometry; for example, the derivation in 1900 B.C. of an algebraic expression for the volume of the frustum of a square pyramid has been described* as the masterpiece of Egyptian geometry.

Two important steps were taken in the improvement of tools. The first was the contriving of a really satisfactory way of fixing the head of a hand-tool to its handle, as in hammers, axes, adzes and so on, by putting the handle through a hole in the head and fastening by wedges—a method which is still in common use to this day. It was comparatively easy to do this as soon as metal tools had been introduced though the earliest metal tools were lashed to the handles in the same way as the stone tools. The second achievement in tools was the lathe, with the bow drill as its probable ancestor. Both did their work of cutting by rotation and a similar drive was used in each case. The innovation was the realization that round shapes could be made by rotating the work and holding a cutting tool against it, the complement of rotating the cutting tool and holding the work stationary as in drilling. The first lathes, crude though they were, represented the first machine tools.

The most surprising and advanced development in metal working

* By Sarton, Ref. 1.

at that time was in the casting of metals by means of the lost wax process, a method that has been revived in the present century for the casting of gas turbine blades.

The engineering progress recorded for this period would have been impossible without the discovery of how to smelt metals from their ores. Copper, bronze and iron tools enabled wood and stone to be worked with an ease and accuracy that could never have been achieved in the Stone Age. The smith began to take his place alongside the woodman in exercising a skill that was essential to the civilization that was developing. Iron implements were a great improvement for farmers, builders and road makers. Iron axes enabled forests to be cleared and iron shears meant that sheep could be sheared instead of being plucked for their wool.

The references that have been made to the crudity of the tools and machines available to the ancients may give rise to the misconception that the quality of the workmanship was invariably inferior to that of the present day. This was not so. By the patient expenditure of skill and time, the craftsman of antiquity produced goods of a quality that has never been excelled and that is displayed in many of the objects taken from the tombs of ancient Egypt and Mesopotamia. The most beautiful ornaments, jewellery and furniture were found that were different from but just as good as the best that we can make today. Their linen was as fine as our products from a modern loom and, likewise, we cannot today with all the aids of modern technology produce any better porcelain than was made in China three thousand years ago. The advance of technology has enabled us to produce a greater variety of goods but they are not always better in style or finish; our greatest achievement has been in vastly reducing costs and labour, so that a much greater volume of production is now attained.

REFERENCES

(for abbreviations see List of Acknowledgments)

1. Sarton, G., *A History of Science*. O.U.P., 1953.
2. Berriman, A. E., *Historical Metrology*. J. M. Dent, London, 1953.
3. Woolley, Sir L., *Excavations at Ur*. Ernest Benn, London, 1954.
4. Chubb, M., *City in the Sand*. Geoffrey Bles, London, 1956.

BIBLIOGRAPHY

Aitchison, L., *A History of Metals*, 2 vols. Macdonald and Evans, London, 1960.

Glanville, S. R. K. (editor), *The Legacy of Egypt*. O.U.P., 1942.

Lucas, A., *Ancient Egyptian Materials and Industries*. Arnold and Co., London, 1934 (3rd edn. 1948).

Montet, P., *Everyday Life in Egypt*. Arnold and Co., London, 1958.

Petrie, Sir W. M. Flinders, *The Arts and Crafts of Ancient Egypt*. T. N. Foulis, London, 1909.

Sarton, G., *A History of Science*. O.U.P., 1953.

Singer, C. (editor), *A History of Technology*, vol. i. O.U.P., 1954.

Talbot Rice, T., *The Scythians*. Thames and Hudson, London, 1957.

Vowles, H. P. and M. W., *The Quest for Power*. Chapman and Hall, London, 1931.

Woolley, Sir L., *Ur: The First Phases*. King Penguin, London, 1946.

Woolley, Sir L., *A Forgotten Kingdom*. Max Parrish, London, 1953.

CHAPTER III

The Greek and Roman Periods, 600 B.C.–A.D. 400

W̶e come now to a period of great development so far as
machinery is concerned. In Egypt and the Mediterranean
the Greek and Roman Empires successively dominated the
scene. The Greeks under Alexander the Great had overthrown the
Persians, who had ruled Egypt from 525 to 323 B.C., and set up a
government, with Ptolemy as King, at Alexandria, which quickly
became a famous centre of learning, attracting scholars from all
over the known world. To the east, civilizations in Persia, India and
China were developing machinery, and from the work of Needham
it is now becoming generally known that several important mechanical
achievements were developed in China many centuries before they
were adopted in the West. In the Mediterranean the study of science
and philosophy flourished under the Greeks in Alexandria and elsewhere for more than four centuries. Every schoolboy has heard the
names of Pythagorus, Euclid, Archimedes, Aristotle and Hero, and is
familiar with some of their achievements. These were but the most
famous of a large band of philosophers and scientists who not only
made discoveries and inventions that we still use today but also left
behind written records of their work, often copied and re-copied, to
inspire the generations that have followed them.

Most of the simple theorems of plane geometry were worked out
and taught by the philosophers of the Hellenic (or Greek) tradition.
Many of them were men who could solve practical problems.
Archimedes, for example, used his knowledge of mechanics to devise
powerful catapults for use in the siege of Syracuse, and he gave a
demonstration of the force obtainable with a multiple pulley by
pulling single handed a fully laden ship along the dry shore. Furthermore, Archimedes was probably, and has often been described as,
one of the greatest mathematicians of all time. He developed the
theory of the lever and he pioneered the science of theoretical

66

mechanics, though much of his work had little practical outcome until centuries later. The Archimedean drill and the Archimedean screw for raising water both bear his name but it is not certain that he was the inventor of either. His studies of floating bodies led him to the discovery of 'the principle of Archimedes'—that bodies immersed in a liquid exert an upthrust equal to the weight of liquid displaced. An important practical application of this principle occurred to him while bathing and pondering over a problem he had been set by the king of Syracuse—how to determine the actual amounts of gold and silver contained in his crown. By immersing the crown in water and measuring its displacement, and the displacement of equal weights of gold and silver he was able to determine the amounts of gold and silver in the crown.

The simple ideas of hydrostatics—buoyancy, centre of gravity and equilibrium were first postulated by Archimedes, and these have remained to this day the most enduring achievement of Greek mechanics. Very little has since been added to the theory of the stability of floating bodies.

The same is not true for the work of Aristotle, whose writings influenced scientific thinkers for many centuries after his death in 322 B.C. for he believed that 'Nature abhorred a vacuum' and went to some pains to prove that a vacuum was a physical impossibility. His ideas on motion were just as wrong. 'If a weight falls from a given height in a given time, twice that weight will fall from the same height in half the time.' It was not until Galileo, one thousand nine hundred years later, showed by observation that bodies of all weights fall from the same height in the same time, that this fallacy was disproved.

It was not Aristotle's fault that after his death, his writings came to be regarded as the ultimate authority on many aspects of science, it was rather the dogma of religion that discouraged men's minds from adventuring with ideas, and from making experiments to find out scientific truths by observation. If the results seemed to challenge the accepted truth of holy scripture, it was wiser not to circulate them.

MATERIALS

Throughout the whole of this period the material most widely used for mechanical engineering purposes was wood. Access to thick forests and the free use of wood was the basis of the economy. Wood was the fuel used for heating, cooking and smelting—except in China where some smelting was done with coal, and wood was the principal material of construction for building furniture, ships, boats, containers, machines and even pipes. The art of seasoning wood was well known to the Greeks and Romans who covered up timber in store with dung to slow up or control the seasoning process. Wood of good quality and in larger sizes than the Egyptian woodworkers had possessed was readily available because the expansion of sea-going trade resulted in timber being brought great distances to be sold to the rich empires of the classical world. Plugging and patching became unnecessary; stopping and coating with gesso—a mixture of glue and plaster—which had been a common practice in Egypt, was discontinued, the trend being aided by the greater precision that could be obtained from using iron tools.

For jointing timber, dowels, tenons and tongues were now being used and less often the dovetail joint known as the 'little axe-head'. Metal nails had come into common use but wood screws remained unknown until the sixteenth century A.D. Thus hinges, locks, and in fact any metalwork attached to wood was always nailed unless wedges would suffice.

Large water pipes of wood were made by the Romans by hollowing out tree trunks and tapering one end of each section to fit into the next. The joints were often strengthened with iron collars.

The qualities of the different varieties of wood were appreciated by the craftsman and each had its special uses. Examples have been found of ash, cedar, cypress, beech, elm, fir, pine, maple and oak. Carpenters and cabinet makers used metal fittings in conjunction with wood, for example, for corner pieces, ferrules, and iron straps for strengthening joints. Wooden spades and plough shares were shod with iron.

The Greeks introduced turnery to woodwork, particularly for furniture. It seems that the accuracy of geometrical form that could be produced so easily on the lathe had a particular appeal to the Greeks. They also, by applying heat, bent wood for use in producing original and artistic forms of chairs and stools.

Iron and steel were used on an increasing scale for tools and weapons. Metal tools became generally available to the farmer and greatly increased the output of the agricultural workers. Iron axes made the clearing of forests much easier and iron or steel was used for most of the simple agricultural implements that we still use today—ploughshares, spades, forks, hoes, scythes, and sickles. Shears were invented for shearing sheep, trimming hair and cutting cloth. The welding of iron by hammering in the forge was discovered in the seventh century B.C., but the Romans made no important advances in iron metallurgy other than the discovery of the tempering of steel already mentioned in Chapter II. Cast iron was unknown to the Greeks and Romans. The Chinese however were making cast iron from the fourth century B.C. onwards and were using it to make

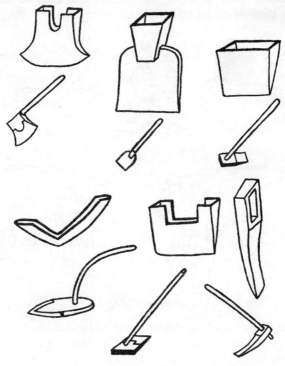

Fig. 19. Chinese tools, 350 B.C. (cast iron).

agricultural implements, moulds for tools and other articles, statuary and weapons. Needham shows some photographs of beautifully preserved cast-iron moulds for chisels or two-pronged hoes dating from the fourth century B.C. and Fig. 19 shows cast-iron tools of the same period.

It is difficult to believe that the Chinese were thirteen centuries in advance of the western world in the manufacture and use of cast iron, but so it was. Needham[1] enumerates five factors which contributed to this lead:

(1) 'The presence in certain regions of iron ores high in phosphorus or minerals containing it which could be added to the charge.

(2) 'The availability (and discovery) of good refractory clays, both for blast-furnace and crucible processes.

(3) 'The use of coal, at least from the +4th B.C. century onwards, perhaps long before, which permitted attainment of high temperatures in the large piles surrounding crucibles. At the same time these would exclude sulphur.

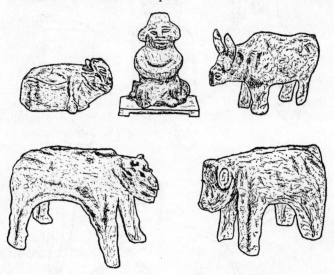

Fig. 20. Chinese statuary, 200 B.C.–A.D. 200.

(4) 'The invention of the double-acting piston-bellows (perhaps in the −4th century, more probably by the +3rd), enabling a strong continuous blast to be employed.

(5) 'The application of water-power to metallurgical bellows as early as the +1st century, greatly strengthening the blast and saving labour.'

The oldest known complex iron casting is a funerary cooking stove about 200 A.D., but during the previous four hundred years there are examples of statuettes and objects in cast iron such as those shown in Fig. 20. Needham has pointed out two very interesting features of the iron and steel technology of China as compared with that of the West; namely that steel was produced from the second century B.C. by the direct decarbonizing of molten cast iron, and that

70

from the fifth century A.D. a great deal of steel was made in China by what he calls the co-fusion process, i.e. the heating of wrought iron and cast iron together, so that the excess carbon in the liquid cast iron passes into the pasty lumps of wrought iron during the process.

The direct decarbonizing of molten cast iron was achieved by subjecting it to a carefully controlled blast of cold air—a method known as the 'hundred refinings', clearly an ancestor of the Bessemer process of blowing air into a converter containing liquid metal (1855) and of the more recent oxygen lancing of steel in which pure oxygen is blown into the liquid metal to reduce the carbon content to the amount desired.

Iron was used extensively in building work, not only in China but (wrought iron) in Greece and Rome. For example, in constructing the Parthenon in Athens (fifth century B.C.) the Greeks used wrought iron beams as cantilevers to support some of the heaviest statues. At another temple at Agrigento (470 B.C.) wrought-iron beams 5 in. × 12 in. and 15 ft. long were used under the architraves and resting on the tops of vertical stone pillars. Iron was also freely used in building for lifting tackle, tongs and clamps. Both the Greeks and the Romans appear to have had considerable faith in the strength of wrought iron for lifting and load-carrying in structures. The possibility of failure of the metal in lifting gear was recognized by Hero, who advised care in choosing the iron, which must be free of folds and fissures. It might break if it was too hard, and bend if it was too soft, thus the load might fall and a workman might be hurt. Some thought for safety evidently existed in those days.[2]

Brass, an alloy of copper and zinc, was the most important alloy produced in classical times. It was made by a cementation process by heating powdered zinc ore with copper and charcoal. Pure metallic zinc was not produced until the sixteenth century A.D. The brass produced by the cementation process was known as white bronze for the difference of chemical composition between brass and bronze was not appreciated at that time. The Romans used brass for coinage, and brass ornaments and fittings were exported to Egypt and other parts of Africa.

Bronze, an alloy of copper and tin, was used extensively for fittings on elaborate buildings of the Roman period, e.g. gilt bronze tiles were used for covering the roof of the Parthenon and some of the churches had bronze pillars and doors. In Pompeii some bronze window frames were found which had been glazed with sheets 28 in. × 21 in. (Glazing of windows had been general practice since the first century B.C.) Brass and bronze were also used for some of the keys (Fig. 21), and door locks devised by the Greeks and Romans, and for pipe fittings (Fig. 22).

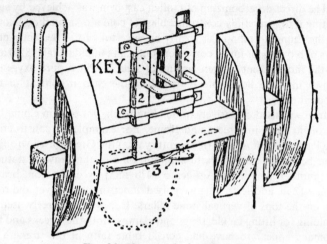

Fig. 21. Homeric lock, 400 B.C.

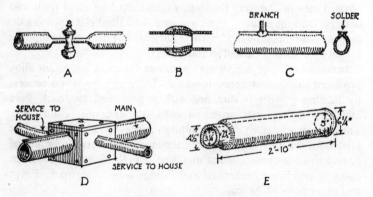

Fig. 22. Roman pipe-fittings.

Lead was produced in great quantities by the Greeks and Romans. It was used to some extent instead of tin in cheap bronze but its major use was in the form of sheet metal as a roofing material, and for making pipes and lining water tanks. Lead solder was well known and similar in composition to that used today. Lead wire was used to draw black lines on parchment and it is from this practice that the term lead pencil has been derived although our present understanding of the term refers to a pencil filled with graphite. Lead was also used extensively for pipes in the Roman water supply system although the Romans were aware of the dangers of lead poisoning. Pipes were made in lengths of ten feet by wrapping the lead sheet round a wooden mandrel, and joined by soldering. The first known example of the standardization of a product used in engineering is that of the lead pipes used by the water authority in Rome, the standard pipe being made of a strip nearly 4 in. wide.[3]

Another engineering material to make its appearance in Roman times was concrete. This was not made from Portland cement (for which the patent was taken out in A.D. 1824) but from pozzolana (a volcanic earth found in quantity in the Alban Hills near Naples) and lime mixed with water and an aggregate of broken brick or stone. The concrete made from it will set under water and resist fire. Mortar made from pozzolana is as strong as the aggregate and it is partly for this reason that so many of the remains of Roman buildings are so well preserved today.

TOOLS

The lathe was the most important new tool to come into general use during this period for it was used extensively by the Greeks and Romans for making furniture, wheel spokes, dishes, and ornaments. No-one knows exactly where, when or how it came to be invented, and no remains of early lathes have survived, though references to the lathe were made by early Greek and Roman authors. One theory is that the first lathe was built in the forest for turning tree trunks, using two standing trees as the head stock and tailstock with the trunk to be turned mounted horizontally between them, a rope wound round the trunk attached to the branch of a young sapling at one end being pulled by a man at the other, so that the trunk rotated first one way and then the other as in the later pole and tread lathe.

Another theory advanced by Reuleaux in 1874 was that the lathe owed its origin to the potter's wheel. He considered that the operation of boring out a hollow cylinder, bowl or dish must have happened long before the reverse operation of turning an external surface. It certainly seems probable that boring out large-diameter dishes could have been done more easily by turning than in other ways. We are left to speculate how the potter's wheel with its vertical spindle came to be adapted for the bowl-turning lathe with its

Fig. 23. Dish lathe.

horizontal axis and running headstock as illustrated by Reuleaux[4] and Klemm (Figs. 23 and 24). This type of machine was used by the Kalmucks as recently as one hundred years ago. Though it is undoubtedly primitive—the spikes were fixed straight into the ground—it is not necessarily ancient.

Fig. 24. Spindle lathe.

74

The same may be said of the spindle lathe illustrated in Fig. 24. This has clearly descended directly from the bow drill or more accurately from the thong drill, since in the illustration the turner needs an assistant to work the thong, but the principle is the same if the turner works the bow with his left hand and the cutting tool with his right. The essential difference from the drill is that the work

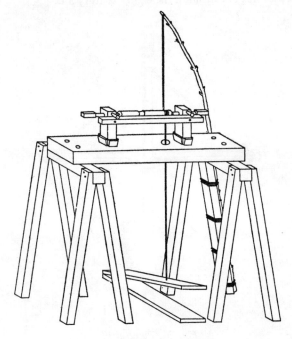

Fig. 25. Pole and tread lathe.

(instead of the tool) is rotated by the bow, and that fixed pivots or centres are provided at each end of the work so that it is constrained to rotate always in the same plane. To make this into a pole lathe, all that is required is a flexible pole and a hinged tread plate, to enable the operator to work alone with both hands free to move and steady the tool. Fig. 25 shows a primitive pole lathe built entirely of wood that was made recently in the author's laboratory. It seems probable that this was the form of lathe most popular from Greek classical times right through to the Middle Ages and still persists for rural crafts such as chair-making in the Chilterns.

Although the bow drill continued to be an essential tool for every carpenter, various improvements to the drill were made about this time. The pump drill or bob drill (Fig. 26) made its appearance and from the many finds of Roman drill bits with square-sectioned shanks it has been suggested that they may have used the brace for drilling. If that surmise is correct it would represent the first use of the crank —otherwise not known until the eighth century A.D.

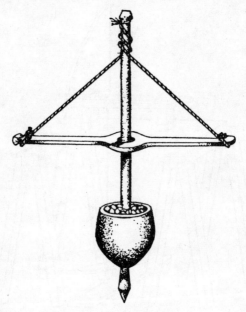

Fig. 26. Pump drill.

Needham[5] asserts that deep drilling in the earth for water, brine and natural gas was practised in China in the first century A.D. but he gives no details of how it was done.

The Greeks made numerous small improvements to the kit of tools used by the carpenter and at least one major improvement was due to the Romans—that of the iron plane for wood work (Fig. 27). The woodworking tools available were made with iron or steel cutting edges and included the auger with its T-handle, the gimlet, the rasp and the draw-knife—forerunner of the spokeshave—and a diversity of improved types of saw including the bow-saw, the frame-saw, the pit-saw and two-handed saws for felling timber. Claw hammers with nail extractors were in use. Other improvements were raked teeth for saws in place of irregular notching, metal ferrules on the handles of chisels and gouges, chisels bevelled on one side of the cutting edge only and spoon bits or quill bits for drilling that have been found on a number of Roman sites in many countries.

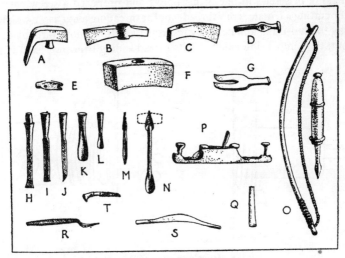

Fig. 27. Roman tools.

One of the important metal-cutting tools—the file—is often mentioned in the classical texts. It was made from the best quality steel from Laconia and was used for cutting, shaping and smoothing metals of all kinds.

The invention of the screw set a new problem for the carpenter. The first screws were almost certainly made of wood, though screws of metal were made in Greece and Rome. To make them a sheet of soft metal was cut in the form of a right-angled triangle having the angle between the hypotenuse and one side equal to the pitch angle of the screw. This soft metal template was then wound around the cylinder on which the screw was to be cut, the templates being held so that one of the arms of the right angle was parallel to the axis of the cylinder. The hypotenuse then traced a spiral on the surface of the cylinder. This method is useful for starting the screw and for checking the pitch at points along the length of a long screw but for cutting a screw with a square thread it can usefully be supplemented by winding around the cylinder two strips of flexible tape—one having the width of the top of the screw thread and the other the width of the channel to be removed in cutting the thread. If these are wound round the cylinder with their edges touching along the whole length of the cylinder and then the one tape removed, the other will serve as a guide for cutting the screw thread. Hero of Alexandria discusses the making of screws in one of his books and proposed a screw cutter for cutting a female screw thread—or for threading a nut as we should say today—using a male screw to propel a cutting tooth through the nut, the screw being guided by its thread sliding over pins fixed in a

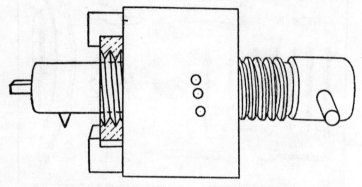

Fig. 28. Hero's screw-cutting device.

frame above the nut. A reproduction of Hero's internal screw-cutting device made by the author is given in Fig. 28.* In the model the nut being screwed was made of transparent plastic material so that the internal screw thread could be seen. Many of the early screws and screw presses were not provided with nuts or female threads but used the device of pins or dowels projecting into a hole larger than the diameter of the top of the male screw to serve in place of a nut.

The screw of the beam press (Fig. 29) used for crushing olives and possibly as a workshop tool for bending wood, was made in this way. We can hazard a guess that the Roman carpenter probably made himself a carpenter's vise with one or two screws very similar to those that we use today.

Ladders with rungs let into the shafts were, as nowadays, used by the Romans both for peaceful purposes, as in building construction,

* First made in 1936 by A. G. Drachmann. *See Journ. Hellenic Studies*, vol. lvi, p. 72.

and for harvesting in orchards and vineyards, and also in warfare for scaling the walls of a fortress during a siege.

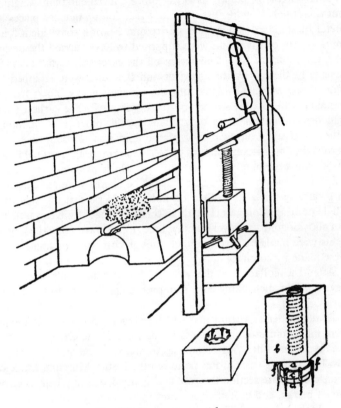

Fig. 29. Beam-press with screw.

The most significant achievement of the Greeks and Romans in metalworking was the development of stamping for coinage. This was the first example of mass production and of the transfer of skill, for in stamping, a few blows by an unskilled worker can produce a finished article of great beauty, for it carries a reproduction of the skill put into it by the artist and the die-sinker. The die-cutting on ancient coins was done with engraving tools similar to those used for engraving gems and seals; that is, small steel cold chisels and gouges that were struck with a heavy hammer to cut and engrave the die. The die itself was a piece of iron or steel.

Coinage is said to have started in the (Greek) cities about the seventh century B.C. when merchants kept precious metals for barter in small ingots or lumps as being more convenient than weighing out quantities of gold dust. The first real coins—that is, pieces of metal of standard weight and purity and bearing some identifying mark—were of silver and were supposed to have carried the design of a turtle on one face. The design on the early coins was carved in reverse in the die by the engraver and the coins were stamped by hammering the metal blank into the die with heavy blows from a hammer while the die was supported on an anvil. The earliest coins had designs on one face only but soon the practice of stamping different designs on each face of the coin was introduced, which involved using a second die as a punch. Constant usage resulted in the upper die becoming worn and convex which explains why many coins are concave or dished in shape. The blanks were cast, usually in groups in a mould so that a number could be cast together at a single pouring. Coins were made in this way of gold, silver, electrum (an alloy of gold and silver) copper, bronze and brass. Only the largest coins were made by casting direct, though cast coins of smaller size were made by counterfeiters.

After the collapse of the Roman Empire the skills that were developed in die-sinking and stamping coinage were lost for centuries.

Draughtsmen's instruments[6] which may have originated in Egypt, were made and used in the classical period and the Greeks ascribed the invention of drawing compasses to mythological gods. There are two 1-ft. folding rules of this period in the British Museum. Made of bronze of square section they are evenly divided along their faces by punch marks, and near the rule joint there is the remains of a catch to hold the rule open. There are also two pairs of compasses in which the hinge pin can be locked in any desired position by a cotter. A set of similar bronze instruments from Pompeii (A.D. 79) are now in the National Museum at Naples. They include rules, compasses, dividers and a pair of outside calipers such as those that can be purchased today.

MACHINES

One of the most interesting mechanical developments of the period was the application of rotary motion to the grinding of corn. Here is an excellent example of necessity being the mother of invention, and as so often has happened in history a number of side inventions resulted that later proved to be of great value in completely different fields.

The importance of the rotary corn grinder, or quern, to the history of mechanical engineering lies in the methods used to turn it. It seems most probable that the hand quern was turned by a crank; when animals were used it was the first application of animal power to industrial purposes; and when water power was applied to it, gearing was used to convert the motion about a horizontal axis of the undershot water wheel to the motion about a vertical axis of the quern.

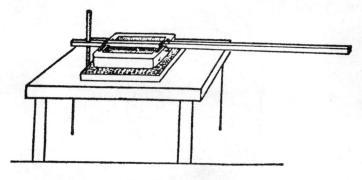

Fig. 30. Pushing mill.

There was an important intermediate step between the saddle-quern (Fig. 4) of ancient Egypt and the rotary quern of the Greeks and Romans. This was the so-called pushing mill (Fig. 30) in which the upper stone was provided with a handle and was often pivoted at one end so that the upper stone moved to and fro in the arc of a circle over the lower stone instead of back and forth in the same straight line as in the saddle quern.

The Roman rotary hand mill was portable and was issued to the soldiers who had to grind their own corn. One portable quern was provided for each five to ten men. The two circular stones were nearly flat, the upper stone being cored out in the centre both to provide a hopper for the corn and to make room for a central supporting pillar fixed in the lower stone, and taking the weight of the upper stone through a spider of wood or metal so that the two stones were only touching each other at the circumference where the ground flour emerged from the crevice between the two stones. Both stones were cut with radial grooves on their working faces.

The upper stone of this type of mill (Fig. 31) could have been twisted with a radial arm to and fro over the lower stone as was the pushing mill, but if it was rotated continuously in the same direction —as it surely must have been—then this represents the first appearance of the cranked handle, a most important event in mechanical engineering, sometimes quoted as having occurred a thousand years later about A.D. 850 (though Needham claims that the Chinese used a crank handle to operate a rotary winnowing machine in the year 40 B.C.). Larger flour mills similar in all essential respects to the Roman quern were used in Western countries for the next two thousand years.

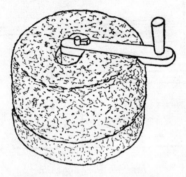

Fig. 31. Rotary hand-mill.

There are two surprising features about the Roman donkey mills. The first is the great thickness of the upper stone which must have been bored out with great labour and internally ground to fit the lower stone and the second is the very close spacing of the donkey mills uncovered at Pompeii (A.D. 79) so that the donkeys operating them were forced to follow a very small turning circle (Fig. 32). A related machine that might be regarded as the ancestor of the edge runner mill was the Greek trapetum (Fig. 33) used for crushing olives during the fifth century B.C. The annular circular basin of this machine was made of lava so that its manufacture was easier than if it had been made of stone. We may wonder if it was machined in a lathe! The operation of the trapetum was effected by pushing the arms D around the central spindle; the upper stones then revolved about the horizontal spindles and at the same time ran round the inside of the basin. This type of motion in which stones revolve about an axis which is itself revolving about another axis must have

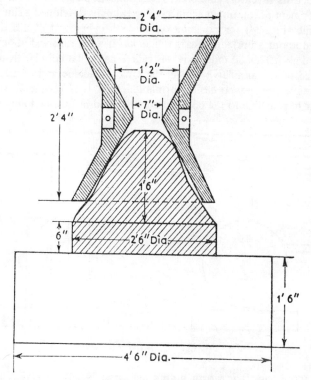

Fig. 32. *Donkey-mill, A.D. 100.*

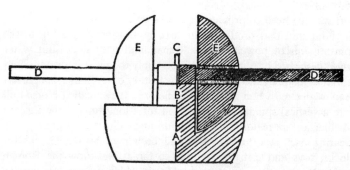

Fig. 33. *Reconstruction of Trapetum from Olynthus.*

intrigued its inventors and marks a significant step forward in the development of rotating machinery. The Romans developed a similar machine (Fig. 34) for crushing olives, with cylindrical stones that rotated about a horizontal axis which was in turn revolved about a vertical axis. This oil mill, as it was called, had a flat circular trough in place of the annular basin of the Greek machine and clearly it must have been much easier to manufacture. It is obviously very similar in principle to the edge runner mill used in Europe today for

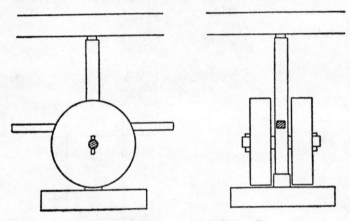

Fig. 34. Reconstructed olive-mill.

mortar mixing. It is worth noting here that Needham claims that the edge runner mill was first invented in China in the first century B.C. and that its use in Europe cannot be fixed precisely until 1607 A.D.[5]

It was the need to provide a suitable drive from a water wheel to the flour mill that produced the first application of gearing to the transmission of power, in the Roman mill with undershot water wheel described by Vitruvius. It is certainly the most important contribution made by the Romans to the history of mechanical engineering.[7] Water mills had been known before—the so-called Norse mill with a vertical spindle had been used for a long time in the Eastern Mediterranean region to drive flour mills directly without gears. Gearing itself was known, and had been used by the Greeks for clocks, toys and instruments, but so far as we know the Roman mill of Vitruvius was the first machine in which gearing was used to transmit power (*see* Fig. 35). The type of gearing used was the lan-

tern gear in which projecting pins on the side of the rim of one wheel
engage with the spaces between bars on the circumference of another
wheel at right angles to the first. Thus a right-angle drive was obtained so that the mill stones could be set with a vertical axis while
the axis of the driving water wheel was horizontal; consequently
much better use could be made of the water supply than had been

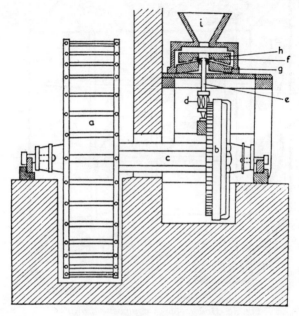

Fig. 35. Vitruvian mill, A.D. 180.

possible with the Norse mill with its vertical-spindled wheel suspended in the millstream. Another advantage of using gearing in this
case was that by suitable choice of the diameters of the two gear
wheels, the speed of rotation of the mill stones could be made greater
or less than the speed of the wheel; also, the speed of the water
wheel while working could be controlled by sluice gates in the
water channel leading to the wheel.

The earliest application of a chain of toothed wheels and pinions
was in the Hodometer, the first kind of taximeter, used on some
chariots and carriages to measure the length of a journey. It will be
seen from the illustration (Fig. 36) that most of the gears are a crude
form of worm and wormwheel. In one sketch the distance travelled

was recorded by the last wheel indexing a hole in a plate, which allowed a stone ball in the upper hopper to fall into a chamber below; in another sketch the distance is shown by a radial pointer operated by the last wheel in the train of gears. We have here the first recording machine or instrument.

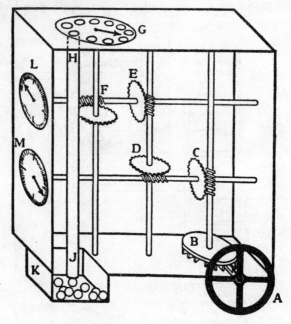

Fig. 36. Hodometer (a composite drawing).

An interesting application of gearing in this era was in the invention of the differential gear which first appeared in the mechanism of the South-seeking chariot in China during the later Han dynasty in the second century A.D. The principle of its operation will be seen from Fig. 37. It should be evident that whichever way the chariot was turned on a smooth surface the figure always pointed in the same direction. According to legend this chariot was used thousands of years earlier, but the date now given is based on the authority of Needham.[5] There is no mention of the differential gear in the Western world during the first millennium A.D.

It is sometimes asserted that it was as a result of the intense study of the lever by the scholars of Alexandria over a period of several centuries that the Greeks attained their understanding of machines.

This is most apparent in the general use of the steelyard for weighing in commerce. The weight of an article was determined not by balancing it with an equal weight at the end of an equal-arm balance, but by sliding a given weight along the arm of the steelyard until the beam was in balance, that is when the moment exerted by the given weight was equal to the moment exerted by the article being weighed.

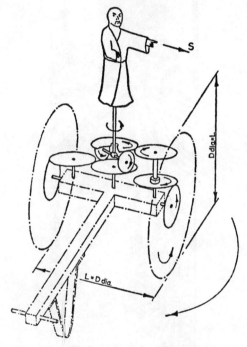

Fig. 37. Diagram of south-seeking chariot.

Such a Roman steelyard was found at Cirencester and is in the museum there. Drachman attributes the steelyard to Archimedes.

Other examples of the lever principle are to be found in the many varieties of the beam press illustrated in Greek and Roman writings. The simplest form was a cantilever beam for pressing grapes or olives; a more elaborate type (Fig. 29) uses a screw at the end of the beam for the pressing operation while a pulley is provided to lift the beam when re-charging the press. One result of the understanding of the beam principle was the ability of Archimedes to calculate the number of pulleys required to lift a given weight with a given force.

Compound and multiple pulleys were used mainly for building construction, the force being applied by a windlass or, in the case of very heavy weights, by a treadmill or 'larger wheel' operated by 'tramping men' or animals. The treadmill is another example of the

Fig. 38. Roman crane and treadmill.

lever principle being used to multiply an applied force. An example of its use to work a crane is shown in Fig. 38, combined with a multiple pulley.

According to Vitruvius the commonest type of crane used for building by the Romans consisted of two masts splayed apart at the ground like modern sheerlegs and supported by guy ropes (Fig. 39).

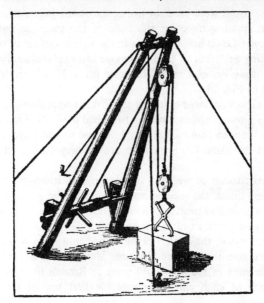

Fig. 39. Roman crane (*Vitruvian*).

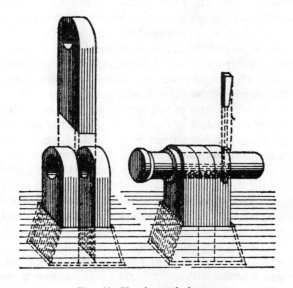

Fig. 40. Hero's triple hanger.

In this case tongs were used for lifting the stone, indentations being cut in the upper surface for the ends of the tongs to enter. For heavier stones Lewis holes were cut in the upper surface of the stone —these being undercut to take triangular lifting plates with a hole in them for lifting or, in a more elaborate form, Hero's triple hanger illustrated in Fig. 40.

For transporting large columns the Greeks and Romans devised several ingenious methods; one was to embed the ends of the columns in circular wooden frames that could be used to roll them along the ground, in the same fashion as we draw heavy cables on a drum today.

The construction of wheeled vehicles greatly improved in Greek and Roman times, the mechanical features of the greatest interest being the wheels and bearings. While solid wheels built up of planks were still used for farm carts, for chariots and other wagons spoked wheels were usual, the spokes being turned in a lathe and mortised or dowelled into the rim and the hub. Metal bands or collars were used at the ends of the hubs, and some in Roman times were lined with bronze at each end—probably the first use of the bronze-bushed bearing.

One of the most puzzling discoveries of this period is the so-called roller bearings in the wheel hubs of the Dejbjerg wagons (first century B.C.) (Fig. 41). The grooves in the hubs are said to have contained rods of wood which turned between hub and axle, but the sides of the grooves would prevent true rolling contact and some skidding of the axle on the rollers would result. There seems to be some doubt as to whether any rods or rollers ever existed since none have been found and so the purpose of the grooves remains a mystery.

The first primitive form of ball-bearing was used by the Romans for turntables on the deck of the pleasure ship of Caligula salvaged from Lake Nemi in modern times by Mussolini. Large bronze balls —presumably castings—were enclosed in a cage and the metal turntable rested on the balls. If the cage was fixed either to the turntable or to the base, as seems likely—then this was not a true ball-bearing since the balls would skid. In any case it is hardly conceivable that means had then been found for making stone or metal balls spherical and all of the same size, an essential requirement for the proper functioning of a ball-bearing.

One of the surprising omissions among the simple machines of the Greeks and Romans is the wheelbarrow. No mention is made of hand carts of any kind and it seems almost certain that the one-wheeled wheelbarrow was not known to Western civilization until the twelfth century, although it was in use in China in the second century. Precisely the same time-lag over the same period occurred in the case of the use of gimbals.

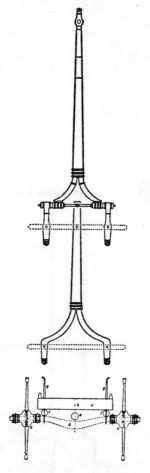

Fig. 41. Dejbjerg wagon, first century B.C.

Some of the missile weapons developed by the Greeks contained new machine elements important to posterity. Among these were the ratchet and pawl used to lock the bow in the stretched position ready for firing the arrow, and elliptical bronze springs which were

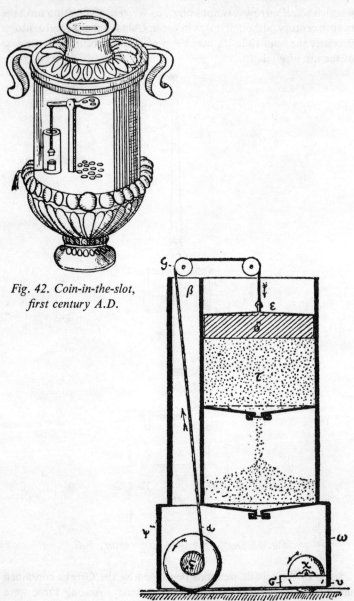

Fig. 42. Coin-in-the-slot, first century A.D.

Fig. 43. Puppet theatre—mechanism.

compressed by the bow arms on pulling back the cord in Philo's chalcotonon. These springs were made of an alloy containing 30 per cent tin which was cast into flat sections that were formed into the elliptical shape by hammering when cold. Another kind of spring that was well known to the Romans was the coil spring which was used in very small sizes for the fastenings on brooches and personal ornaments. Many ingenious clasps and fastenings are to be seen in Roman jewellery.

The Greeks made many mechanical toys and devices to promote wonder and astonishment among worshippers in the temples. These included a coin-in-the-slot machine for dispensing a libation of holy water (Fig. 42) and numerous devices for mechanically-operated puppet theatres (Fig. 43). An interesting feature of the puppet theatre was its driving mechanism, consisting of a heavy lead weight resting on the top of a bed of millet or mustard seed contained in a vertical cylinder or box with a small aperture at the bottom. As the seed ran out of the box the weight fell slowly and by a rope attached to it drove the mechanism; an interesting example of driving by weight.

FLUID MACHINES

In the study of fluid behaviour the Greeks made some enduring contributions e.g. the principle of Archimedes', of which the Romans applied the knowledge to a most remarkable degree in their aqueducts and water supply systems.[3] Open channels were used in preference to pipes, probably because the material for open channels—masonry and mortar—was more easily available, but where pipes were necessary, lead or earthenware were most usual, though pipes of copper, bronze and hollowed stone were also used. The lead pipes were made in standard lengths—not less than 10 ft. long—from cast sheets rolled up with butted edges and welded by pouring molten metal along the joint. Recent testing of some of the original Roman pipes has shown that the longitudinal welded joints were stronger than the wall of the pipe. Stop cocks and valves were seldom used, for it was the practice to allow the water to run continuously through the pipes. The consumer paid a flat rate for a continuous discharge which was estimated—inaccurately—from the size of his supply pipe. It is recorded that sometimes dishonest consumers hammered out their supply pipes when made of lead—to obtain a greater flow, so the water authority was obliged to fit a standardized brass tube to each consumer's supply pipe at the inlet end. The

Roman engineers believed that the quantity of water supplied in this way depended only upon the area of the supply pipe—ignoring altogether the effect of change of head or velocity on the quantity flowing—so that they were never able to obtain a consistent balance of the quantities flowing in various parts of the supply system. Discrepancies were attributed to leakage losses and to the fraudulent practices of consumers. No attempt was made to keep the velocity of flow the same along the length of the aqueducts, for their slope sometimes varied from 1 in 2,000 to 1 in 250 along the length of the

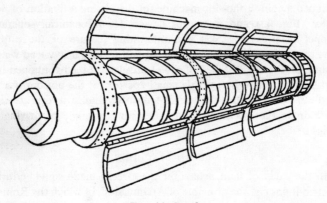

Fig. 44. Snail.

same conduit. They realized of course that a downward slope was needed to secure a flow as the Greeks and Egyptians had done before them.

Some examples of Roman pipe fittings and junctions are shown in Fig. 22. A great step forward was taken during this time in methods of raising water, by devising means for doing it continuously instead of intermittently, as had been the case with earlier methods, for example, the shaduf, or buckets. The two principal methods that came into use were the Archimedean screw or snail, and the chain of pots, both still in use among primitive peoples. Water lifting was of great importance in mining, for irrigation (it was claimed that the whole of the Nile delta was irrigated at this time by Archimedean water screws), for watering gardens, for dissolving salt in open salt pans, and for public and domestic water supply.

The Archimedean water screw (Fig. 44) in its simplest form consists of a coil of pipe that is rotated with its axis slightly inclined to the horizontal while the lower end is below water level. The centrifugal force induced by rotation increases the rate of flow and of course requires a greater expenditure of energy. An axle supporting the coil was provided through the centre with one bearing below the water and the upper bearing above the water level. For a small inclination the 'snail' could be operated by slaves using their feet, making it a kind of treadmill; alternatively it could be turned by a proper treadmill of the cage wheel type through gearing. It is possible to 'screw' water upwards with a spiral rotating inside a stationary cylindrical pipe. The Romans made a unit of this kind for mining that was virtually standardized. It had a wooden core on which was wound a spiral of wood or copper and was covered with a wooden casing so that in operation the water was forced up the annulus between the core and the casing by the rotation. It was pivoted at each end and was set at an angle of from 20–45° to the horizontal according to requirements. 'Snails', as they are called, were usually used for lifting water through small heights and when they were used in deep mines a great number were operated in series. For large heights the chain of pots, or Persian wheel, was more usual.

The Persian wheel was a wheel with a horizontal axis, the wheel carrying a rope chain to which earthenware pots or containers were attached at regular intervals. The lower end of the chain carrying the pots (Fig. 45) dipped into the water, and to facilitate filling the pots were sometimes provided with a small hole in the bottom to allow air to escape as the pots entered the water. A trough was provided at the top into which the pots emptied themselves as they passed over the wheel.

A related water-raising device, the square pallet chain pump, was used in the first century in China, though its Western prototype, the chain and rag pump, did not appear until about the seventeenth century. In these pumps one side of the chain passed up through a vertical conduit and the water was drawn up the conduit by the pallets or rags sealing the conduit as they entered it at the bottom.

In Egypt the Persian wheel and the snail were used extensively in early classical times, but during the Roman period the former was replaced to some extent by the Noria or Egyptian wheel—a large wheel constructed of wood with a horizontal axis, its lower extremity dipping below the water surface. Pots or buckets were attached to the rim of this wheel, which was rotated by a treadmill or by animal power through gearing, or even by paddles attached to the rim of the wheel when there was sufficient velocity of flow in the river from which the water was being raised.

The Romans used a series of eight pairs of such wheels in series in the second century A.D. for draining the copper mines at Rio Tinto in Spain. Each wheel was about 15 ft. diameter giving a lift of about

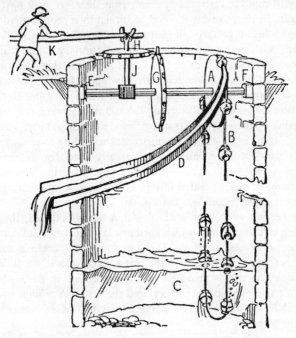

Fig. 45. Chain of pots.

11 ft. per wheel, a total lift of 88 ft. It is believed that the wheels were turned by treadmills.

Experimenting with wheels for raising water gave rise to the invention of water wheels for producing power and driving machinery. It has been established that the oldest type of water wheel (or water mill as it is called, since all the early water wheels were used for driving mills to grind corn), was the Greek water mill which later spread north all over Europe and became known—as it is to this day—as the Norse mill (Fig. 46). It had a vertical spindle carrying a set of radial vanes or paddles at its lower end which was let down into the running stream. It was most effective if combined with a mill race and chute to direct the water flow onto one side of the wheel. It was limited to fast-running streams which mainly occur in hilly regions and was suitable for small quantities of water moving at high velocities, consequently its power output was small, less than ½ h.p., which was little more than that required to operate a donkey mill. The Norse mill was simple to construct and could be worked with a small rapid stream, so that it remained popular for grinding small

CUTAWAY VIEW OF NORSE MILL
SHOWING PADDLE CLEAR OF WATER

Fig. 46. Norse mill.

quantities of corn in out-of-the-way places in Europe and Asia where suitable water supply was to hand. Furthermore, it could be built very cheaply and required little maintenance. Mills of this kind were still in use until recently in the Shetland Islands and the Faeroe Islands, though superseded elsewhere from the eighth century onwards by the Roman water mill. They can be looked upon as the original ancestors of the modern water turbine.

The horizontal water mill of Vitruvius was the most important contribution of the Romans to the history of mechanical engineering —apart from its reaction on social life by lessening the daily labour

D

of grinding corn, though it was not universally adopted in Rome until the advent of Christianity led to the freeing of the slaves. Its importance to power transmission has already been discussed. Now we should direct attention to the complete change in the power output of machinery that was achieved by this means. It has been estimated that in its most primitive form as much as 3 horse-power was developed by this machine, making it far the most powerful prime mover then in existence—nor was that all, for when it had been fully developed some centuries later, as much as 50 horse-power was

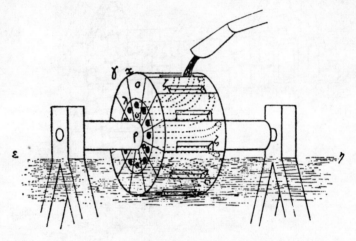

Fig. 47. Philo's whistling device.

generated by machines that were similar to the original in all essential respects. The important elements of the Vitruvian water mill were the undershot water wheel with a horizontal axis, and the right-angle gear drive to the mill stone which rotated about a vertical axis. The water was directed on to the paddles at the bottom of the wheel by the mill race and chute, so that the wheel was able to make use only of the kinetic energy—or velocity—of the water directed onto it.

The best example of a Vitruvian mill that remains is at Venafro, near Naples, where not only a mill race and the remains of an aqueduct have been found, but also an imprint of the water wheel— which was about 6 ft. in diameter and is estimated to have generated about 3 horse-power.

The earliest evidence of the idea of an overshot wheel appeared in one of the whistling devices of Philo (Fig. 47). It consisted of a drum made of wood or copper mounted horizontally, the lower part being below the level of the water. The drum was divided internally into compartments and was rotated by a stream of water from a chute above the periphery of the wheel. As the drum rotated air was trapped in the chambers and later expelled by the water through tubes which had a whistle at the end. The apparatus had no practical utility but

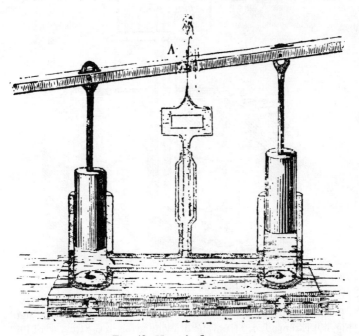

Fig. 48. Heron's fire pump.

may have led to the development of the overshot water wheel at the end of this period of history.

Many of the remaining mechanical devices of this period that are important to the history of mechanical engineering depend for their action on the piston and cylinder combination. When this was first used is unknown but it was probably used in the form of the syringe for mummification in Egypt and by the second century B.C. if not before, in the fire engine described in Hero's *Spiritalia* (see Fig. 48). The automatic suction and delivery valves (not shown), were either thin metal plates with metal pins as guides or of the stem and mushroom type. It will be noticed that there are no connecting rods or cranks, the pistons being solid with the piston rods whose top ends were coupled to the rocking beam for hand operation. The beam was loosely connected at its junctions with the piston rods. It is stated explicitly in Hero's book that this machine was used as a fire extinguisher.

Fig. 49. Ctesebian suction- and force-pump.

Force pumps were also used for domestic purposes. The figure (Fig. 49) is reproduced from the drawing given in the treatise of Philo of Byzantium and shows two pumps for forcing water up from a well, the delivery pipes being 15 ft. long. The suction valves were the plate type with four pins as guides but the delivery valves would appear to have been different—possibly of the leather flap variety that were used in bellows. Provision was made for sliding motion between the operating levers and the tops of the piston rods.

Well-known examples of Roman plunger pumps were discovered at Silchester and similar bronze pumps found at Bolsena, Etruria, are now in the British Museum.

The remains of the Silchester pump are shown in Fig. 50. They were found in 1895 during the excavations of the site of the Roman city of Silchester in Hampshire, and consist of a block of oak 22-in. long that has been bored out to form the receiver, a rising pipe, two pump barrels and the connecting passages for a two-throw pump. The receiver, rising pipe and pump barrels were all lined with lead $\frac{3}{16}$ in. thick. The bore of the pump barrels was about 3 in. and that of the delivery passages 2 in., the receiver 5 in. and the rising pipe

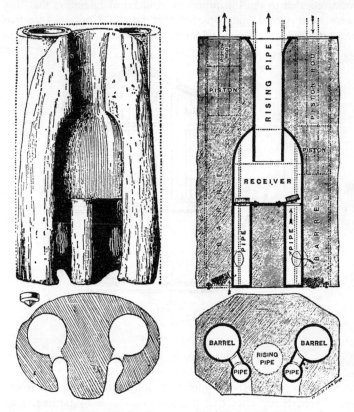

Fig. 50. Roman pump from Silchester.

$2\frac{1}{2}$ in. A reconstruction of this pump is shown in the same figure. Leather flap valves to which lead weights were attached by riveting are shown, as one of these lead weights was found with the remains.

The two bronze pumps found at Bolsena are similar in principle and were probably parts of two Roman fire engines, the larger of these having barrels about $1\frac{1}{2}$ in. diameter. The disc valves were riveted loosely to the seats and were free to flap open or close as the pistons moved within the cylinder barrels. The pistons were hollow and about 2½ in. long, and the stroke was apparently about 3 in. The pumps were partially submerged in water when in use so that they were not required to produce a suction but only to force the water upwards. The remains were examined in 1911 by J. P. Maginnis, who reported to the Institution of Mechanical Engineers that they were so encrusted and corroded that he could not be sure that the parts had been turned and bored but he thought that the excellent

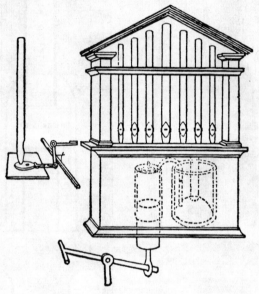

Fig. 51. Hero's hydraulic organ.

fit of the spigot and socket joining the horizontal pipe where it was sawn through suggested that this had been made by turning. That being so, it is extremely likely that the pistons and cylinder barrels were turned and bored in a lathe, and, furthermore, both Vitruvius and Hero in their writings about this type of pump say that the cylinders were bored and the pistons smoothly turned and rubbed with oil.

All the five piston pumps just mentioned were for pumping water but the same type of piston pump was used by the Greeks and Romans for pumping air. Air from these pumps was used to operate Greek toys and spectacular devices used in temples and places of amusement. Notable among them was the water organ (Fig. 51) so called because slightly compressed air was stored under water at nearly constant pressure in a miniature diving bell in a water tank situated between the air pump and the keys operated by the organ player. Organs blown by leather bellows had been known before, but the 'water organ' attracted attention presumably because the 'diving bell' provided a better reservoir of air under pressure than

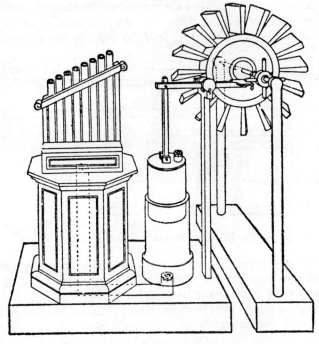

Fig. 52. Hero's wind organ.

was available in the bellows type. A variation of the water organ of particular interest to mechanical engineers is one described by Hero in which there was no keyboard and the piston pump was operated by a windmill, so that when the wind blew all the organ pipes would sound (Fig. 52). The vertical piston was loaded with weights to cause it to descend and was lifted by a crude form of cam and rocker worked by the horizontal windmill. Both the windmill and the cam and rocker came to be developed for more useful purposes during the centuries that followed.

This was the first reference to any sort of windmill by the classical writers, though it is believed that wind-driven prayer wheels were in use at this time in the Far East. Dr. Needham states that the rotary fan was being used in China for ventilation, a rotary winnowing machine being driven by a crank handle and that blowing engines driven by water power were being used in China.

One of the most exciting inventions of the Chinese took place in the fourth century B.C. when the piston-type double-acting bellows was invented for blowing furnaces in the production of cast iron. The writer has had a model made of this with one side transparent

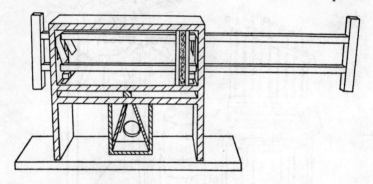

Fig. 53. Chinese double-acting blower.

so that the construction can be clearly seen (Fig. 53). Flap valves for suction were required at each side of the piston but a single delivery valve at the centre served for both sides by flapping on to either of two valve seats from its vertical position. The blower was made square or rectangular in section—the full-size piston being as large as 4 ft. by 4 ft.—and was operated by a single man pushing and pulling from one end only. A characteristically Chinese detail is that the piston packing was of chicken's feathers—as in the model. The importance of this device is that a double-acting machine was in use as early as 400 B.C. in China and that its use continued for nearly two thousand years before it was applied in the West in the sixteenth century.

HEAT ENGINES

Heat engines were not applied to useful purposes during this period but several interesting devices were developed as toys that were later recognized as illustrating most important principles in the development of power from heat. The first of these was Hero's famous aeolipile (Fig. 54). This was a hollow metal sphere filled with steam through trunnion pipes that were fixed at their lower ends to the lid of a cauldron placed over a fire. As the water boiled in the cauldron steam issued in two jets through tangential outlet pipes on the sphere, turning it around on its trunnions by the reaction force of the steam jets.

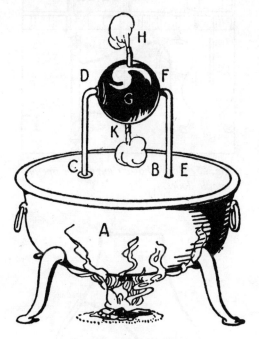

Fig. 54. Hero's aeolipile.

Hero also described a device which may be termed the first hot-air engine (Fig. 55). Its purpose was to close the doors of a temple without human agency, by the heat from a fire lit on the altar in the temple. The working parts were all below floor level and could not be seen by those in the temple. The doors were mounted on spindles having extensions down into the crypt below, round which ropes were wound, one of each being fastened to a bucket that would become filled with water as the altar fire became hot, and thereby opened the doors. The altar was made of metal, hollow and air-tight so that the air

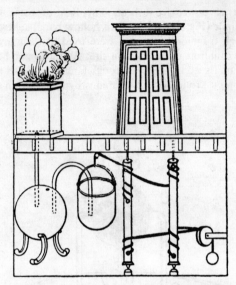

Fig. 55. Hero's hot-air engine for opening temple doors.

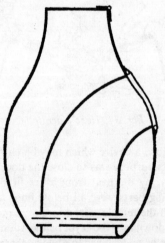

Fig. 56. Roman water-heater.

within was forced down a pipe into a vessel below, where it displaced water into the bucket. When the fire went out, the air in the altar cooled, its pressure was reduced and water was syphoned back from the bucket into the container. We have here an example of a hot-air engine using a syphon.

The third example is a water heater used by the Romans (Fig. 56). It was made in the form of a jug and contained an internal charcoal fire resting on water tubes passing from one side of the vessel to the other. Although this device was for water-heating rather than for steam-raising, it may be regarded as the ancestor of the modern water-tube boiler.

REVIEW

The mechanical engineer will be most impressed by the ingenuity of the Greeks in devising machines during this period. Here it seems the leisure and curiosity of the philosophers and scientists was used to good effect in producing such elements as the screw, the ratchet and pawl, and gearing and machines like the force pump, the water wheel and Hero's aeolipile. Some of these were elaborated to mystify or amuse, either as toys, or in the theatre. When Rome conquered Greece and ruled the Mediterranean, these devices were assessed and useful ones adopted, for the Romans were not themselves inventive but they organized and standardized, improving on many of the Greek ideas and adapting them for further uses.

The study of machines by the Greeks and the Romans was an accomplishment rather than a 'professional activity'. Its object was to instruct and amuse rather than to lessen the toil of human beings, hence the pre-occupation of the Greek philosophers with puppets and mechanical devices that could be used in the theatre—the windlass and the pulley, that could let down a god from above the stage—a Deus ex Machina—so useful in the Greek drama.

The application of science to the study and classification of machines was a great achievement. Hero of Alexandria described what he called the only five simple machines by which a weight can be moved by an applied force—the lever, the wedge, the pulley, the wheel and axle, and the screw. This proposition was accepted for centuries and was even elaborated into the idea that all more complicated machines consisted of these constituent elements, combined in various ways. Two of the most important machines described by Hero—the aeolipile, or primitive reaction steam turbine, and the hot-air engine for opening and closing the doors of a temple—may fairly be called the first heat engines, but the former was devised as a toy and the latter to inspire wonder in the temple worshippers so that the tremendous practical possibilities of the heat engine as a prime mover passed unnoticed for the next fifteen centuries. A similar fate befell the first primitive type of gas turbine—the zoetrope—a lamp cover caused to revolve by the rising current of hot air—invented in China about A.D. 180.

Though the possibility of using the heat engine to do work was missed, it was in this period that animals were first harnessed to machinery for purposes other than transport and tilling the soil. The Roman donkey mill was used for grinding flour, and the treadmill or cage wheel, operated either by men or animals, was used by the Greeks and the Romans for operating cranes, pumping water, and for a variety of other purposes. More important still for mechanical engineering was the invention of the water wheel (about 400 B.C.) and eight hundred years later the harnessing of the wind to drive prayer wheels in China, from which windmills probably evolved.

There was a greater variety of materials that could be used for building machines; not only iron, but brass, bronze and lead were used increasingly, though by far the greatest part of mechanical construction was still done in wood, of which little has survived. The lathe, the screw press, the pump and the water wheel were made almost entirely of wood and so were the gear teeth of the Vitruvian mill; likewise the Chinese piston bellows, which is doubly important because of its part in producing cast iron and because it was the first piston machine in which both sides of the piston were used, i.e. it was double-acting.

However, in spite of the wider range of materials and devices available it is extremely doubtful whether the skill of the craftsman in using his resources was any greater than it had been a thousand years before. Indeed in many crafts it was certainly less; for example, in the making of fine jewellery, textiles and furniture, the ability to produce exquisite workmanship with fine design and finish seems to have been lost when the ancient and decaying Egyptian empires were overcome by the Greeks.

Needham has drawn attention to the surprising achievements of the Chinese who, before the end of this period, were using mechanical inventions which did not appear in the West for another nine to fifteen centuries. They included iron-casting, the wheelbarrow, the rotary winnowing machine with crank handle, the edge runner mill, double-acting piston bellows and the rotary fan. It is surprising that the ideas incorporated in these machines did not diffuse thence to the West, for trade with China (via Persia or Egypt) particularly in silks and pottery, dates from pre-Roman times.

Among all the many mechanical achievements of this period pride of place must be given to the water wheel, in so far as it was the first prime mover to be invented. It was the first self-acting machine to do useful mechanical work without the aid of men or animals, and although very crude in its original form, it marked the beginning of a new era for human civilization. At the same time theoretical advances were being made in the field of fluid mechanics by that intellectual giant among engineers Archimedes, who was then making his studies of buoyancy, the centre of gravity and the equilibrium of floating bodies. All these have endured to this day as the first steps in the science of hydrostatics.

In marked contrast we have two principal heat engines devised during this period, Hero's hot-air engine for opening temple doors, and his whirling aeolipile, 'an interesting device for amusement and instruction'. Both were extremely sophisticated machines invented much before their time, inasmuch as the scientific basis needed to explain their operation was completely lacking. Both involved the construction of metal vessels that could withstand heat and pressure. These limitations explain why it was nearly two thousand years before they were understood and applied to engineering work.

While it is generally conceded that Western civilization owes a tremendous debt to Greece and Rome in the fields of art, literature, philosophy and politics, it is arguable that science and engineering owe them little for the reason that the classical tradition regarded the practice of all useful work as being intellectually inferior to occupations that produced no practical result at all, such as the discussion of the meaning of words or of abstract mathematical propositions. This attitude of regarding classics and the arts as superior to sciences, and particularly applied sciences, is something we have inherited from classical times and it has persisted in the United Kingdom up to the present day, with the result that many of the ablest minds in the community have never been willing to adjust themselves to the changes which stemmed from the industrial revolution and to play the part they might have done in the development of our culture.

REFERENCES

(for abbreviations see List of Acknowledgments)

1. Needham, J., *The Development of Iron and Steel Technology in China*. Newc. Soc., London, 1958.
2. Singer, C. (editor), *A History of Technology*, vol ii, p. 753. O.U.P., 1956.
3. Ashby, T., *The Aqueducts of Rome*. O.U.P., 1935.
4. Reuleaux, F., *The Kinematics of Machinery*. Macmillan, London, 1876.
5. Needham, J., *Science and Civilization in China*. O.U.P., 1954.
6. Dickinson, H. W., *T.N.S.*, xxvii, 1949–51, p. 73.
7. Elton, J. and Bennett, R., *History of Corn Milling*, vol. ii. Simpkin Marshall, London, 1899.

BIBLIOGRAPHY

Moritz, L. A., *Grain Mills and Flour in Classical Antiquity*. O.U.P., 1958.

Needham, J., *Science and Civilization in China*. O.U.P., 1954.

Quennell, M. and C. H. B., *Everyday things in Ancient Greece*, 2nd edn. Batsford, London, 1954.

Quennell, M., *Everyday things in Homeric Greece*. Batsford, London, 1954.

Singer, C. (editor), *A History of Technology*, vol. ii. O.U.P., 1956.

Talbot Rice, T., *The Scythians*. Thames and Hudson, London, 1957.

CHAPTER IV

The Dark Ages and the Renaissance
A.D. 400–A.D. 1500

The eleven centuries which followed the fall of the Roman Empire can be divided for our purpose into two distinct parts of nearly the same length: the six centuries from A.D. 400 to 1000, the so-called Dark Ages in western Europe, when much of the heritage from Greece and Rome was forgotten; and the remaining five centuries from A.D. 1000 to 1500, which saw the birth of technology in the West where a number of primary inventions were made that were essential to the progress that followed.

The decline and fall of the Roman Empire did not occur suddenly, but the power and protection which the Pax Romana had provided for so many centuries over such a vast area of the world was gradually withdrawn. Accordingly the states which had enjoyed this protection had to learn to defend themselves, to rule themselves and to develop the resources and skills of their own peoples. This took time, for the more warlike peoples such as the Danes, the Vikings, the Huns and the Goths raided their neighbours and dissipated their resources in the fighting which prevailed over much of the area of the West during the Dark Ages.

The cultural and religious centre of the Roman Empire had been moved from Rome to Byzantium (Constantinople, now Istanbul) in A.D. 330 by the Emperor Constantine, and here Roman rule and culture persisted at least in name for another seven hundred years, but Byzantium, though a centre of art and commerce, never became a centre of learning in philosophy and science in the way that Alexandria had been. Much that remained lay unrecognized and unused so that, for example, the texts of Euclid and Aristotle were lost for some centuries until Latin translations appeared in the thirteenth century.

During the Dark Ages in Western Europe public works of all kinds

111

were neglected though farmers retained their skills in agriculture and craftsmen were still sought by kings, barons and religious orders.

The Christian Church was steadily growing in power after the fall of Rome and it had a great influence on the development of engineering during this time. Whereas its teaching was against personal ostentation the Church helped to keep alive many of the luxury arts and crafts for church furniture, and also commissioned the building of great churches, monasteries and cathedrals throughout Europe. Moreover, it was the church that popularized the arabic numerical system in almanacs and calendars, since arithmetic was used in those days chiefly for calculating the dates of religious observances. Until the thirteenth century all but the simplest reckoning was done on the abacus—a collection of beads sliding on wires mounted in a wooden frame. With this device addition and subtraction were easy, but not the other arithmetical processes. Our present familiarity with the so-called arabic system of numerals makes it hard for us to appreciate how difficult and inconvenient it must have been to make any complicated computation with the Roman notation, particularly with large numbers.

At the end of the twelfth century Leonardo of Pisa, a merchant, learnt from the Arabs the Indian system of numbers, which makes the rules of arithmetic very simple. This is due to the value of a digit being dependent upon its place in a line of digits, as in the number 1,111, where each has a different value according to its place in the line. Leonardo of Pisa wrote a book explaining the system for technical and commercial use for which it quickly replaced the old system. We still refer to it as the arabic system though it originated in India.

An example of the extent to which the new number system aided commerce can be gleaned from the story of the Merchant of Prato (near Florence), who left his estate to his home town in 1410. When rediscovered in 1870, his papers included one hundred and fifty thousand business letters, five hundred account books, four hundred insurance policies, three hundred deeds of partnership and thousands of bills of lading, bills of exchange and cheques.

The Church also encouraged primary education and fostered national languages and literature by arranging for the Gospels to be translated and sending missionaries who knew the native language, such as Saint Patrick to Ireland. Travel abroad was always easy and frequent among the religious orders. Perhaps the most spectacular contribution of the Church to the development of mechanics was in the stimulus it exerted in the invention of the mechanical clock.[1]

Lewis Mumford has described how the orderly daily life of the church and the monastery led to the need for time measurement and eventually to the development of the clock.

Monastic orders flourished, the number of monasteries in the order of Benedictines being 40,000 at one time, and they became so rich that they have been called the founders of modern capitalism. They kept many of the engineering crafts alive within their walls and adopted and improved many processes. They introduced water power for flour mills, drained fens, made cloth and were important wool producers. They copied manuscripts and wrote new works: for example, a Cistercian monk, Villard de Honnecourt (1270) was an engineer who drew a book of plans for all kinds of water-driven machinery. A good deal of our knowledge of daily life of this time comes from such records as the Lindisfarne Gospel, A.D. 700, and the works of the Venerable Bede, A.D. 673–735.

During the Dark Ages men were more concerned to preserve facts that had been collected in classical times than to attempt to make new discoveries and interpretations, but nevertheless five hundred years later (A.D. 1500) the most advanced western countries were superior in their mechanical achievements to any earlier society.

The primary factors which helped to produce these achievements were the scientific method, with its laboratory for experiment, critical observation and measurement; the university, for study and learning, where knowledge was organized on an international basis (the earliest universities were Bologna, 1100, Paris, 1150, Oxford, 1200, Cambridge, 1229 and Salamanca, 1243); the printed book, which included the inventions of paper, the printing press and movable type and extended the range of communication of facts and ideas, giving permanence to the written record; and the mechanical clock, which provided its own standardized product—minutes and seconds following each other regardless of human events. Furthermore, the clock was also the first automatic machine in general use and the men who made clocks were later available to apply their skill to other useful mechanical arts.

Of equal importance at this time was the growth of the use of mechanical power. Slavery had become unpopular in most Christian countries as contrary to the teachings of Christianity, and the ravages of the Black Death brought scarcity of labour in its train as it swept across Europe from Asia later in this period. The growth of power other than man-power helped to fill this need. The adoption of the rigid stuffed collar and iron shoes for horses released more power for transport and for horse-driven machines, while the invention of the multiple yoke for oxen in the tenth century was another improvement to the same end. Water power was more readily tapped by the introduction of the overshot wheel and wind-power was harnessed in windmills. It seems probable that the increased use of power transformed life in the Middle Ages as much as the steam engine was to do later on during the industrial revolution.

The earliest western scientists of this period were much preoccupied with the practical application of their knowledge, as can be seen from the remarkable prophecy of Roger Bacon who, in the thirteenth century, foresaw 'ships without rowers, chariots without animals that could move with unbelievable rapidity, flying machines and unheard of engines'.

Still more remarkable, both in his vision and his achievements, was that giant among mechanical engineers—Leonardo da Vinci. Engineers are apt to think of Leonardo in terms of the mechanical inventions attributed to him such as his bobbin, spindle and flyer for reeling yarn, the centrifugal pump, breech-loading cannon, rifled fire-arms, anti-friction roller bearings, universal joints, the conical screw, rope and belt drives, link chains, bevel gears, spiral gears, the submarine and the parachute. Many of these were not made during his lifetime and had little influence on the practice of mechanical engineering until centuries later. What was probably much more important was his exposition of the experimental method, and his studies in mechanics, statics, dynamics and hydraulics and the pioneer work on aviation made in his notes 'On the flight of birds'. In the field of hydraulics, for example, many mechanical engineers are unaware that it was Leonardo da Vinci who five centuries ago was the first to describe the profiles of free jets of water, the formation of eddies at abrupt expansions, the interference of waves and the hydraulic jump. He also proposed the streamlining of bodies immersed in a fluid to reduce friction and made observations of the flow of water by means of suspended particles in a glass-walled channel.[2]

THE INVENTION OF PRINTING

The invention of printing with movable type took place during the last century of this period and it has been described as the pre-eminent mechanical contribution of the Renaissance because it had the greatest influence on the social development of Western civilization.[3] The success achieved in printing with movable type was due to the bringing together of a number of technical developments in different fields such as the manufacture of paper, the right quality of ink, the art of the engraver in making hard metal punches for the letters, the development of type founding (Fig. 57) (a form of die-

Fig. 57. The type-founder.

casting) and the adaptation of the screw press for the purpose of printing (Fig. 58).

The manufacture of paper is another example of the length of time taken for a particular technique to travel across the world. Paper-making was invented in China about A.D. 100 and Usher gives the following dates for its manufacture elsewhere: Samarkand, 751, Baghdad, 793, Egypt, 900, Morocco, 1100, Spain, 1150, France, 1189, Italy, 1276, Germany, 1320.

Books were first printed with movable type in Germany about 1440 but printing was taken up most vigorously in the succeeding years in Italy for it is recorded* that by A.D. 1500 the number of printing presses in use in Europe was 1,050, their locations being: Italy, 532, Germany, 214, France, 147, Spain, 71, Holland, 40, others, 46. Thirteen of these last were in England, including the celebrated Caxton Press. There were 7 in London, 3 in Westminster, 2 in Oxford and 1 at St. Albans.

Fig. 58. Amman's printing press.

The mechanical skills involved in developing the process were as follows: hard metal punches were engraved by the methods that had been developed for coinage and jewellery, each punch having engraved at one end a mirror image of a single letter. (The punches were first made in brass or copper but later iron was used.) The punches were struck with a hand hammer to produce depressions in copper,

* Brit. Museum Cat., 1908, Proctor and Pollard.

thus making a mould or matrix that was an exact reproduction of the letter to be printed. The type was made by casting molten type metal (a composition of tin, lead and antimony) into the mould which was made adjustable in width to provide for varying widths of type. Each type casting, carrying the imprint of a single letter at one end, was then removed from its mould, the runner sawn off and filed to the required length so that all the letters would stand at the same height when placed on end in a tray. The letters were arranged in lines in the tray by the compositor who had to secure them tightly with wedges so that their location was fixed accurately in both directions in the

Fig. 59. Roman coin (sestertius of Vespasian commemorating capture of Jerusalem, A.D. 70).

plane of the tray. The assembled type was then wiped over with ink, covered with a sheet of paper and placed in the screw press so that a vertical pressure could be applied over the whole surface of the paper where the type had been inked. The pressure was then released by the screw and the new printed sheet of paper peeled off the surface of the type.

It would seem that the most important engineering achievement in this sequence of processes was the type-casting—for casting molten metal into a metal mould made by stamping was surely new. All the other processes involved were well known at the time, particularly engraving and reproducing the pattern by stamping or hammering the engraved tool onto copper, silver or gold. This was the regular occupation of those who made coins, some of which had designs with fine lettering (*see* Fig. 59). The screw press was in common use at that time for many purposes, e.g. for crushing olives and grapes and even for 'ironing' pieces of damp cloth.

The use of this type of mould for type-casting had other important consequences, the chief being that the square stem of the type thus made was sufficiently accurate for the letters to be interchangeable— the first example of interchangeability in manufacture. Without this accuracy the method would not have been successful. Other features of the process were the accuracy that had to be achieved by the engraver in positioning his work correctly on the end of the punch, so that the resulting printing was in line, and also the jig, which must have been used for locating the punch when striking the mould, and the skill required to strike the punches evenly so that the letters were all of the same depth. These difficulties were finally solved four hundred years later when type-founding became completely mechanized.

MATERIALS

In the field of materials the two most significant developments of the time were the manufacture of large castings, particularly of iron, and the birth towards the end of the period of the study of the science of strength of materials.

Wood remained as before the principal material of construction for machines and their supporting structures, though the larger powers being developed by water wheels and windmills involved the use of massive sections—sometimes whole tree trunks for shafts and their supports. During this period a beginning was made in the use of coal for smelting.

The invention of gunpowder, said to have been brought to the West from China, led to the need for strong metal tubes before explosives could be used to propel and project missiles in warfare. During the fourteenth century a number of gunsmiths set up foundries to serve this need, particularly in the Rhineland, in Germany, where many famous gun-masters used the same techniques for casting cannon in time of war and for founding church bells in times of peace. The making of cast bronze cannon has much in common with the making of large bronze church bells, though soon the bronze for cannon gave place to iron. The achievements of those early German gunsmiths have been summarized by Professor Matschoss in his book entitled *Great Engineers*: for example, in 1404 a giant cannon 11½ ft. long and 4½ tons in weight was constructed in Austria.

The greatest technical achievement of the Middle Ages was the production of cast iron. We have seen that this had long been practised in China where the double-acting piston type of blower was used to supply the air blast.[4] In Europe leather bellows with wooden top and bottom boards and leather flap valves were used, and by the use of water wheels sufficient power was available to operate very large sizes of bellows so that the necessary high temperature to melt the iron could be obtained (*see* Fig. 60). Casting into moulds of stone

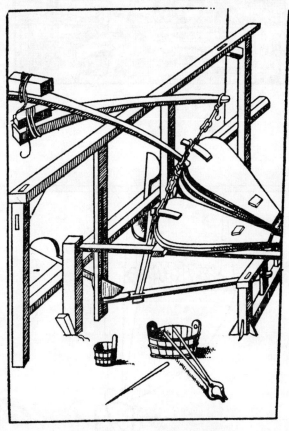

Fig. 60. Bellows worked by cams.

and clay was superseded by sand-casting in divided mould boxes during the thirteenth to fourteenth centuries. The use of water power contributed to the iron industry in other ways, in crushing ore and in operating tilt hammers for forging and stamping, in turning the grinding wheels of the armourers and in wire drawing using a crank (Fig. 61).

Fig. 61. Wire-drawing using a crank.

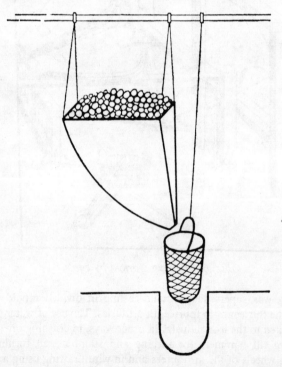

Fig. 62. Tensile test of wire (Leonardo da Vinci).

The first tensile testing machine on record is a sketch of apparatus for testing the strength of wire, in the notebooks of Leonardo da Vinci (Fig. 62). A basket or container is attached to the wire to be tested.[5] During the test fine sand is fed from the hopper shown through a fine hole which is closed by the spring when the wire breaks. When this happens the basket falls only a short distance and comes to rest without overturning. Leonardo emphasized the need to repeat the test several times so as to check the result. He also studied the strength of beams and concluded that for a beam of uniform cross section 'the part that is farthest from the supports will bend the most' and that the strength of beams supported at both ends varies inversely as the length and directly as the width. His study of the strength of columns led him to state that this varies inversely as their lengths and directly as some ratio of their cross section.

It appears that Leonardo's understanding of the thrust produced by an arch was sound and that he was able to use correctly the principle of virtual displacements to analyse systems of pulleys and levers used for hoisting.

TOOLS

During the disorganization of the Dark Ages the activity of the craftsman was restricted and it is not until the ninth or tenth century or even later that we begin to learn of any improvements in tools and machines; for example, the wood plane, which was well known to the Romans, was not reintroduced to the Western world until the twelfth century and was not in common use until the thirteenth century.

The new hand tool of the greatest importance in this period was the brace. It provided not only a rapid means of drilling medium-sized holes in wood, but it represents for the mechanical engineer one of the first uses of the complete crank, i.e. with four right-angle bends, the connecting rod being supplied by the arm and wrist of the operator. Other improved tools were developed for drilling such as the tree-trunk borer (Fig. 63) and the beam drill, the latter providing means for exerting very large pressure on the drill for drilling metal. The reciprocating saw continued to be used as the principal means of cutting soft stone or wood into pieces, as in the making of cubes and planks. Mumford says that in 1322 a saw mill operated by water power was in operation at Augsberg, but whether by cams and levers or by connecting rod and crank is not stated, but there were some in

England in 1376. By the end of this period the lathe was certainly operated by a foot treadle and crank, particularly for small work.

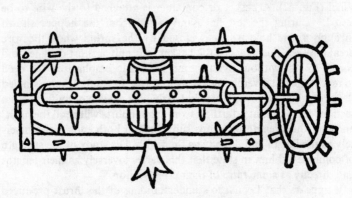

Fig. 63. Tree-trunk borer.

In the notes of Leonardo da Vinci there is a sketch of a treadle lathe with a flywheel, having one fixed centre and one running centre (Fig. 64). This was in advance of its time in two respects, the fly-

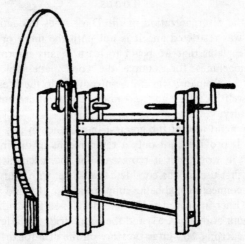

Fig. 64. Lathe (Leonardo da Vinci).

wheel which gave more uniform motion than would occur without it, and the treadle and crank which enabled a single operator to rotate the work continuously in the same direction. Contemporary

Fig. 65. Pole and tread lathe with bowl.

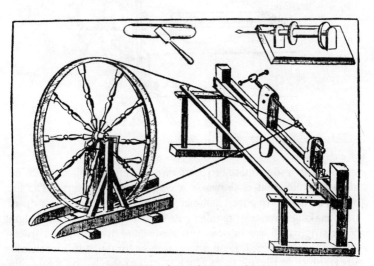

Fig. 66. Lathe with great wheel.

illustrations of the two types of lathe in use in the middle ages by furniture makers show the work being directly driven between two fixed centres as in ancient times. For small work the furniture maker used a pole and tread lathe (Fig. 65) and for heavy work a lathe driven by the great wheel (Fig. 66), the same combination that was used by George Stephenson in building his first locomotive at the beginning of the nineteenth century.

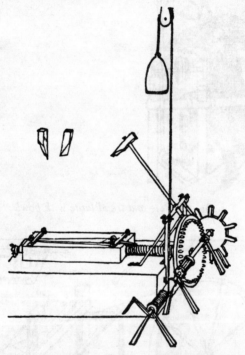

Fig. 67. File-cutter (Leonardo da Vinci).

Leonardo also proposed to mechanize the highly skilled operation of cutting files that is shown in a contemporary picture.[6] The file maker used a sharp-edged hammer with which he struck the strip of steel, making a series of parallel cuts close together down the length of the strip which was subsequently heated and quenched for hardening and then tempered. Leonardo's machine for doing this mechanically is shown in Fig. 67, from which it can be seen that by means of a screw, a cam, and gearing, the cutting tool was caused to fall automatically when the file had been moved forward the required distance between the cuts.

A model of Leonardo's screw-cutting machine is shown in Fig. 68 and can be seen at the Science Museum, South Kensington. It is similar in principle to Hero's device for cutting internal threads (Fig. 28) but in this machine two male threads are used as masters and the pitch of the thread cut on the rod in the centre can be

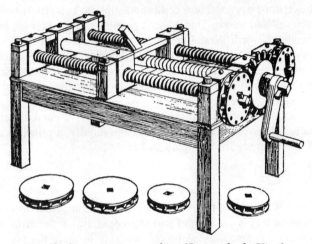

Fig. 68. Screw-cutting machine (Leonardo da Vinci).

altered by varying the gear ratios as well as by altering the pitch of the threads on the two master screws.

The first iron sewing needles were made in the fourteenth century in Germany. They had a closed hook at one end in place of an eye. Iron sewing needles with an eye began to be made in the Low Countries before A.D. 1500. Iron nails were being made at that time by hand, by driving small pieces of metal through a succession of graded holes in a plate.

The making of coins was in abeyance for some centuries after the fall of the Roman empire so that the skills involved had to be re-learnt in the Middle Ages when their manufacture was resumed. The new coins were thinner, simpler and in lower relief, some being struck on one face only. Coins of the eighth and ninth centuries were crude by comparison with those of Roman times. The dies for these were not engraved but made by punches which could make straight lines, triangles and curves from which the design was composed. Gradually the practice of engraving the dies to produce more elaborate designs was restored until by the fifteenth century it had become usual.

In the field of agriculture a most important and ingenious improvement was made to the plough by the addition of the mould board. This device, which comprised a curved, concave surface of wood or metal attached to the back of the ploughshare, was used to guide and turn over the heavy soil in a continuous ribbon, completely burying the top soil and weeds. This made it possible to turn over the heavy clay soils of North West Europe in which the ploughs of classical times would have been virtually useless. The Roman plough had been little more than a wedge which in the dry soils of the Mediterranean shores sufficed to push the soil aside into ridges after the land had been roughly cleared by harrowing.

The mould board was the first application of a curved inclined plane to fold over a strip of material continuously, a principle since used in machinery for wrapping and folding.

MACHINES

Towards the end of the period the use of machines had become so widespread that it can be said that the mechanical era of civilization had begun. The sources of power available to drive the machines were water, wind, and the muscular energy of men and animals. Thus the initial stages of the industrial revolution took place before the development of steam power had started, and the mechanical inventions of classical times were put to use on a scale that had never been known before. Hero's five simple machines—the lever, the wheel and axle, the pulley, the wedge and the screw were all incorporated in more elegant and elaborate machines in which power was used to do mechanical work of various kinds much faster than it could be done by the primitive methods used in earlier civilizations. Nor was that all, for the understanding of what could be done with falling weights, ropes, pulleys and gear wheels suitably combined with accurate workmanship, led to that triumph of the mechanician —the mechanical clock.

Mumford has described its origin to meet the needs of the regular communal life of the monasteries of the Middle Ages.[1] He regards the mechanical clock as the foremost machine of modern technics, the key machine of the modern industrial age—principally for two reasons: its automatic action, and its own special product, accurate timing, without which our modern industrial civilization would be unthinkable.

What appears to have been an important step in the evolution of the mechanical escapement has been suggested recently by Derek J. Price.[7] In the Chinese Astronomical Clock Tower of Su

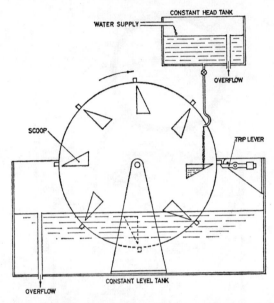

Fig. 69. Model of Su Sung escapement, A.D. 1090. (Simplified.)

Sung, A.D. 1090, there was a slowly revolving wheel carrying scoops on its periphery each being filled in turn by dripping water. The wheel moved intermittently, being held in check by a trip lever and weight while each scoop was being filled by the dripping water. When the weight of water in the scoop was sufficient to trip a weighted lever, the wheel moved onwards until it came to rest, when the next scoop had reached the filling position. An important feature of the device was that the scoops in their lowest position were below the water level, so that the energy in the wheel at the moment of release by the trip lever was absorbed by the water during the time that the wheel was moving into position for filling the next scoop. Overflow pipes to keep constant levels were required both in the supply tank and for the outlet, in order that the device would keep time. Fig. 69 shows the essential features of a working model made recently in the author's laboratory.

The primary problem of the mechanical clock was to control the rate of fall of a weight attached to a cord wound round a drum, so that the drum would rotate slowly at constant speed as the weight fell. Something of the same sort was done in the mechanism of the

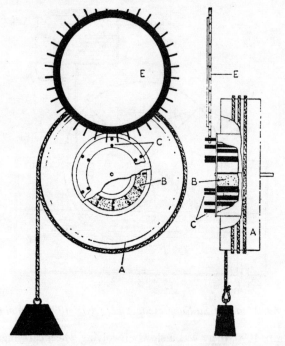

Fig. 70. Alfonso's mercury clock.

Greek puppet theatre (*see* Fig. 43) where the falling weight rested on the top of a quantity of millet seed contained in a canister so that the rate of fall of the weight was controlled by the rate at which the millet seed was falling through an aperture at the bottom of the canister. Such an arrangement was not sufficiently controllable or slow enough for a mechanical clock, but a somewhat similar idea was used in the mercury clock of Alfonso X described in a book compiled by Spanish savants in 1276. In this clock the rotation of the drum to which the falling weight was attached, was controlled by mercury contained in annular partitions within the drum (Fig. 70). As the weight fell the rotation of the drum lifted the mercury, so counterbalancing the driving weight, whose rate of descent was determined by the rate at which the mercury trickled through holes in the annular partitions of the drum.[8]

This arrangement did not come into general use because it was

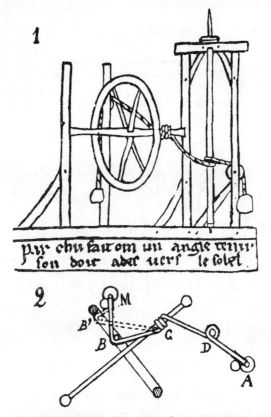

Fig. 71. Villard de Honnecourt's rope escapement.

superseded before the end of the thirteenth century by a simple device to interrupt the action of the falling weight, so that the pointer indicating the passage of time was moved forward in steps instead of continuously. The device is called the escapement and it is the most essential element in all mechanical clocks and watches. How it came to be invented is obscure. Many believe that it had its origin in a primitive device described in the album of Villard de Honnecourt about 1250 (Fig. 71). This consisted of a flywheel with a horizontal axle mounted in pivots and having a rope wound round the axle. Each end of the rope was passed over a pulley and kept in tension by a weight. The weights at the two ends were unequal. If the flywheel was out of balance then its oscillation could be caused to grip the rope as it moved one way and to slip on the return, so that the heavier

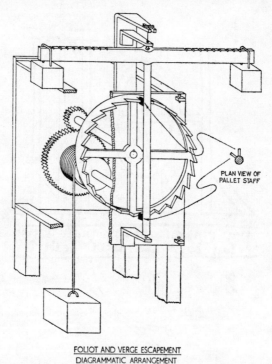

PLAN VIEW OF
PALLET STAFF

FOLIOT AND VERGE ESCAPEMENT
DIAGRAMMATIC ARRANGEMENT

Fig. 72. Foliot balance and verge.

weight fell, lifting the lighter one. The rope escapement was not a satisfactory or reliable device however, and it is doubtful if it was ever used in a practical clock.

The first practical mechanical escapement was the foliot balance and verge, the essentials of which are illustrated in Fig. 72. The first mechanical clocks were made with this escapement during the latter half of the fourteenth century; the date of origin is discussed fully by Usher who concludes that earlier references to clocks refer to water clocks.[3] The verge escapement was an essential feature of all the many mechanical clocks used during the following three hundred years. They comprised tower clocks which indicated the hours both by an hour hand and by chimes, cathedral clocks, sometimes operating puppets as at Wells Cathedral, and domestic clocks. The essential features of the verge and foliot balance escapement are the suspended crossbar or foliot with two small cursor weights at its ends. Fixed to its centre immediately below the thread suspension is the vertical arbor spindle or verge which carries two projections or pallets which engage alternately with the teeth of the crown wheel so that under the action of the driving weight the crown wheel rotates half a tooth forward for each swing of the foliot balance. The fourteenth-century clock installed in Salisbury Cathedral and the Dover Castle clock (probably fifteenth century) were originally both of this type. The wheels and frames for these clocks were of wrought iron, the framework being secured by cotters. No timber was used in the construction and there were no screw fastenings. These early clocks were probably among the first machines to be made entirely

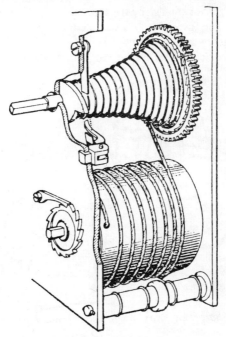

Fig. 73. The fusee.

of metal and they set a standard of accuracy and workmanship which remained in the lead for centuries.

The early clocks were made by blacksmiths, who were used to working on mill gears. Gradually as his work became more precise the clockmaker came to occupy a place of esteem in the community similar to that of the toolmaker centuries later. Clockmaking became so highly developed in Germany in the fifteenth century that domestic clocks were more common there than elsewhere in Europe, and clocks made to indicate minutes and seconds were first made there.

The imperfections of the gearing used in clockwork stimulated the mathematical study of the most suitable shape for gear teeth. A mathematical analysis of the cycloidal curve was published in 1451 and the epicycloid was discovered in 1525.

The earliest watches were made before 1500, and they are usually referred to as the 'Nuremberg eggs'.* Two important inventions were involved. The first replaced the falling weight by a coiled spring within a drum, and the second was the fusee, a device to maintain the same force to the train of gear wheels as the coil spring was unwound although the force supplied by the latter was diminishing as it unwound itself. The fusee (Fig. 73) was a conical pulley connected to the spring drum by a strong gut cord wound around the circumference of each, the conical pulley having ridges to prevent the gut from

Fig. 74. Sharpening sword on grindstone, ninth century A.D.

* Though they were thus miscalled through faulty translation and probably did not originate in Nuremberg, but Milan.

slipping axially along the pulley. As the coil spring unwound, the gut was wound off the conical pulley onto the drum which was of uniform diameter and thus the diminishing tension in the gut operated on the gear train at an increased radius, and provided a uniform torque or twist to the gear train if the angle of the cone had been correctly chosen. Leonardo shows a number of sketches of the fusee in his notes but the idea was well known before his time.

Fig. 75. Cranks and connecting-rods.

Perhaps as important to the mechanical engineer as the development of the clock was the use made during this period of the crank —first the hand-operated crank, as for a grindstone (Fig. 74) or a spinning wheel, and later the crank combined with a connecting rod which we have seen illustrated in Leonardo's treadle lathe (*see* Fig. 64). A manuscript of 1430 shows a mill operated by two cranks and connecting rods coupled to treadles (Fig. 75). Another example of obtaining motion in a straight line by the use of a crank worked by

Fig. 76. Spinning by hand.

Fig. 77. Spindle wheel.

a water wheel is shown in Fig. 61 where a man sits on a swing using tongs to enable the crank to draw the heavy iron wire through a draw plate.

The invention of the spinning wheel during the fourteenth century resulted in the mechanization of the most tedious operation in the production of textiles. The eleventh-century illustration, Fig. 76,

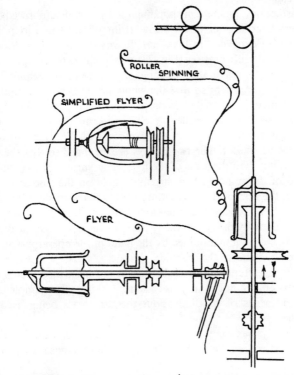

ROLLER SPINNING

SIMPLIFIED FLYER

FLYER

Fig. 78. Development of the spindle.

shows spinning by hand, and the next illustration, Fig. 77, shows the process mechanized by the use of the spinning wheel in the fourteenth century. In hand spinning the spindle is vertical but in the mechanized version the spindle is mounted horizontally in bearings rotated quickly by a rope drive connecting it with the large wheel which is turned by one hand of the spinner. The process was in both cases intermittent and consisted of two distinct operations—spinning and winding-on. The spinning operation comprises both twisting and drawing out the fibres of the yarn. The spinning wheel presents an important step in the progress of mechanical technology since it is among the earliest examples of the use of continuous rotary motion. The flywheel effect of the large pulley was also an advantage, and it is among the earliest examples of the use of a rope drive between large and small pulleys, the rope being crossed to minimize slipping at the small pulley.

A further development in spinning, the flyer, which followed about a century later, is often erroneously attributed to Leonardo, possibly because he sketched it in his notes. It was a well-known device by 1480 when the illustration (Fig. 78) was published. The purpose of the flyer is to make the spinning and winding-on simultaneous instead of intermittent so that the whole process can be continuous. This is achieved by having two projecting arms mounted on a hollow spindle with a driving pulley at one end, and the spool mounted loosely on a spindle between the two arms of the flyer and driven by a separate pulley. The two pulleys are arranged so that the spool rotates faster than the arms of the flyer. The yarn is passed through the end of the spindle, then through a hole on the side and over a hook on the arm of the flyer before it is attached to the spool. Thus the three basic processes can be carried on simultaneously; the winding on is done by the spool, the twisting by the flyer, and the drawing out by the tension maintained by the hand of the operator who feeds the yarn by hand to the machine.

The flyer is important both because it has been used ever since in all forms of spinning, and also because it is an early example of two different speeds of rotation about the same axis being used in a machine.

During the next century a treadle was fixed to move the large wheel by foot, so leaving the operator free to use both hands for feeding the yarn.

FLUID MACHINES

The technical revolution which began in the twelfth and thirteenth centuries was based very largely on the many applications of the water mill which was the prime mover, supplying the power to carry out many operations previously done by freemen or slaves. In this the monasteries took the lead, for they were generally self-supporting and by using water power and mechanizing arduous tasks that required human labour the monks had more time for prayer and meditation—surely the best reason that has ever been advanced for adopting mechanization! The Cistercian monks in particular made a practice of building their monasteries near rivers where a suitable water supply could be used to operate the machinery. The extent of this machinery can be gathered from a description of Clairvaux Abbey in France where it was said that the river, on entering the abbey buildings, first ground the corn and worked the sieves which separated the flour from the bran; then heated the water for the beer for drinking; then operated the fulling machines in a building where cloth was made; then it was used in the tannery where footwear was produced; and finally it carried away all the refuse. Evidence of the extensive use of water power in a monastery during the twelfth century can also be seen from the ruins of Fountains Abbey in Yorkshire.

Fig. 79. Bellows water-pump (Vegetius).

The water wheel had been improved, since Roman times, by the Christian monks and also by the Moslems. These improvements—not only in the wheels themselves but also in the machines they operated such as large bellows for smelting, stamp mills and fulling mills—were among the more important technical ideas brought back to Western Europe by the Crusaders.

A large number of illustrations of water wheels towards the end of this period depict overshot wheels (*see* Fig. 109) whereas the Roman wheels—like that of the Vitruvian mill (*see* Fig. 35) were undershot. An interesting variation of the undershot water wheel was the floating mill first used on the river Tiber when Rome was besieged by the Goths in 537 (Fig. 80). The idea spread during the centuries that followed and there are records of floating mills on the Tigris during the tenth century, in France on the Seine, the Garonne and the Loire during the twelfth, thirteenth and fourteenth centuries and on the Thames at London Bridge early in the sixteenth century. However, the most primitive form of water wheel—the Greek or Norse mill (*see* Fig. 46)—continued in use not only until 1500 but in inaccessible regions such as the Hebrides until quite recent times. Though it is slow and inefficient, requiring large volumes of water moving at high velocity, it has the virtue of cheapness and simplicity, for little special skill and no expensive material is required for its construction.

Fig. 80. Floating mill, Armenia.

The number of water mills of all kinds increased greatly from the fourth to the fifteenth century. For example, Forbes states that in the Aube district in France there were 14 mills in the eleventh century, 60 in the twelfth and 200 in the thirteenth. A remarkable increase in the number of water mills in use occurred in Britain where according to the Doomsday book (1086) there were more than five thousand water mills in the southern half of England. Most of these mills had undershot water wheels. Before 1500 it was usual for every village of reasonable size to have a water mill, often belonging as a monopoly to the lord of the manor or the local abbey who charged tolls for grinding the corn.[9] They remained substantially the same for a long period but improved methods of building, balancing and grooving the millstones were gradually evolved.

Although it was not until after A.D. 1500 that great interest was shown in water pumps, there must have been considerable use made of piston pumps of various kinds during the Middle Ages. Ewbank asserts that there are numerous proofs in old authors that pumps were in common use in wells during the fifteenth century.[10] In its simplest form the piston pump was made of wood, either square or circular without any valve, the water being drawn up below the piston, which fitted tightly in the cylinder.

The bilge or burr pump was used on board ship. It had one valve in the bottom and a conical leather piston which could collapse and open out on successive strokes of the piston. Ships' carpenters were expected not only to repair but to make pumps during a voyage if the need arose.

At the very beginning of this period, A.D. 400, travellers in China observed the use of prayer wheels rotated by the wind, sails mounted on a vertical axis being used. Windmills of this type, that is, with sails fastened to radial arms moving around a vertical axis, were used in the Far East for grinding corn some centuries later, for in the thirteenth century a traveller described a two-story windmill of this kind in which the grindstones were housed in the upper part, the lower one containing the sails being rotated by the wind which entered through trumpet-shaped ducts in the walls of the structure. This remarkable description of what we might now call an enclosed air turbine has a considerable resemblance to the Greek or Norse water mill, the only significant differences being that the turbine rotor was much larger, with the blades made of fabric sails instead of wood, and that the turbine rotor was enclosed in a casing whereas in the Norse mill the rotor was immersed directly in the water stream.

Windmills in western Europe at this time were the orthodox type of post mill illustrated in Fig. 81. Here the sails were in the open and moved about a horizontal axis driving the upper millstone through gearing causing the millstone to move about a vertical axis. The machinery was contained in a large wooden box which was mounted on a heavy vertical post so that the whole mill could be moved round by a long lever to face the sails into the wind.

Probably the great effort required to turn this post mill led to its being partly superseded in the fourteenth century by the tower mill.

Here the machinery was contained in a massive tower, often circular in section and constructed of stone or brick work, and only the cap of the tower—usually of wood—was rotated to bring the sails into the wind. No details of construction are available for windmills before A.D. 1500.

It was in the Low Countries of Europe that the post and tower mills

Fig. 81. Post mill.

were used in greater numbers than elsewhere, though it became as common a form of prime mover as the water wheel in many other countries. Many pictures of the Low Countries show great numbers of windmills within the same area, not only for grinding corn but for pumping water, particularly in connection with the drainage of the polders or fens. Towards the end of this period the Dutch had become famous as millwrights and they were in demand as builders of windmills and also as drainage experts all over Europe.

REVIEW

As we look back at these achievements of more than a thousand years, we can see that the most important events affecting the history of mechanical engineering were the discovery and improvement of the scientific method with its techniques of observation, recording, analysis and experiment, and the establishment of universities where men learnt these methods and were encouraged to add their contribution to the common stock of knowledge. Furthermore, the West acquired from the East during this period and adapted to its own advantage the art of printing and the arabic system of numerals, both probably essential factors in the rise of science and engineering that followed so rapidly.

In mechanical engineering itself the mechanical clock and the use of the crank to convert reciprocating motion to (or from) rotary motion were both of great importance. Perhaps no less important was the building of more powerful machines to perform the tasks previously done by men and animals, and we perceive for the first time attempts being made to improve the output and efficiency of machines. More power was obtained from horses as draft animals by giving them better harness and metal horse shoes, while the motion of rotating wheels was used, not only to measure the passage of time, but for spinning, and for drawing out and twisting threads for textiles with 'the flyer' which used the differential speeds of two wheels, rotating about the same axis, to such good effect.

The discovery of how to make heavy metal castings, first applied in the West to the making of guns and church bells, was later to be applied to more useful purposes in the making of machinery and structures. The principal improvements in tools were in the plough and the lathe. Improvements to the lathe were of several kinds; it was made smaller and more accurate for clockmaking, which introduced into engineering ideas of precision of workmanship that had not been known before; secondly, by rotating the work continuously in the same direction the cutting speed was more than doubled as compared with the earlier lathes, where the work was turning first one way round and then the other. In consequence larger pieces could be machined in the lathe and the turning of metal became more general when more power became available, either by driving the lathe from a treadmill or water wheel, or by means of a crossed belt from the great wheel— a heavy flywheel turned by men operating two cranked handles. The use of the great wheel to drive line shafting continued for the next five to six centuries, persisting until the first half of the twentieth century, for driving rotary cleaners for hair brushes in barbers' shops, for example.

The appearance of the crank was a most important event in the history of mechanical engineering. Still more so was the combination of the crank, treadle and flywheel shown by Leonardo da Vinci in his sketch of a treadle lathe. Here we have both the crank and connecting rod working together to transform reciprocating motion into rotary motion and the use of the inertia of the flywheel to continue the motion past the 'dead' centre of the crank. Both of these important conceptions have been used in machinery ever since.

In Leonardo da Vinci's notebooks we see amongst many other stimulating ideas the first gropings towards the applied science of strength of materials.

If we had to choose the most important from among these achievements surely it would be the invention of movable type for printing. Though printing itself came from the East, the invention of movable type in Germany was the outstanding technical triumph of the West in this period. Apart from its tremendous social consequences, it was the first example of an engineering process in which interchangeable parts, high accuracy and the transfer of skill to produce many identical products were the features essential for success.

Next in importance was the mechanical clock. Here for the first time was a machine that produced a new product that could not be made by hand. Accurate time was a commodity so prized by everyone that to possess a watch gradually became a symbol of success, and it is probable that no other machine has had such a profound effect upon the way of life of civilized mankind. To the mechanical engineer of today the clock signalizes the first automatic machine in which small amounts of energy can be stored up to be released under controlled conditions over long periods of time without the interference of any external agency. It was, of course, the first of a long line of automatic machines that have become more precise and more ubiquitous right up to the present day.

REFERENCES

(for abbreviations see List of Acknowledgments)

1. Mumford, Lewis, *Technics and Civilization*. G. Routledge and Sons, London, 1934.
2. Hart, I. B., *The Mechanical Investigations of Leonardo da Vinci*. Chapman and Hall, London, 1925.
3. Usher, A. P., *A History of Mechanical Inventions*. Harvard U.P., 1954.
4. Needham, J., *Science and Civilization in China*. O.U.P., 1954.

5. Timoshenko, S. P., *History of Strength of Materials*. McGraw Hill, London, 1953.
6. Singer, C. (editor), *A History of Technology*, vol. ii. O.U.P., 1956.
7. Price, D. J., 'On the origin of clockwork, perpetual motion devices and the compass', *U.S. Nat. Mus. Bull.* 218, p. 81. Smithsonian Institution, Washington, D.C., 1959.
8. Singer, C. (editor), *A History of Technology*, vol. iii. O.U.P., 1957.
9. Elton, J., and Bennett, J. R., *History of Corn Milling*, vol. ii. Simpkin Marshall, London, 1899.
10. Ewbank, T., *Hydraulic Machines*. Tilt and Bogue, London, 1841.

BIBLIOGRAPHY

Berriman, A. E., *Historical Metrology*. J. M. Dent, London, 1953.

Crombie, A. C., 'Augustine to Galileo', *The History of Science A.D. 400–1650*. Falcon Press, London, 1952.

Forbes, R. J., *Man the Maker*. Constable, London, 1950.

Gomme, A. A., *Patents of Invention*, publ. for British Council by Longmans Green, London, 1946.

Hart, I. B., *The Great Engineers*. Methuen, London, 1928.

Hart, I. B., *The Mechanical Investigations of Leonardo da Vinci*. Chapman and Hall, London, 1925.

Klemm, F., *A History of Western Technology*, trans. by D. Singer. Allen and Unwin, London, 1959.

Lilley, S., *Men, Machines and History*. Cobbett Press, London, 1948.

Matschoss, C., *Great Engineers*, trans. by H. S. Hatfield. George Bell, London, 1939.

MacCurdy, E. (editor), *The Notebooks of Leonardo da Vinci*, 2 vols. Reprint Society, London, 1938.

Needham, J., Wang Ling and D. J. Price, *Heavenly Clockwork*. C.U.P., 1960.

Origo, Iris, *The Merchant of Prato*. Jonathan Cape, London, 1957.

Sarton, G., *A History of Science*. O.U.P., 1953.

Singer, C. (editor), *A History of Technology*, vols. ii and iii. O.U.P., 1956, 1957.

Usher, A. P., *A History of Mechanical Inventions*. Harvard U.P., 1954.

Vowles, H. P., and M. W., *The Quest for Power*. Chapman and Hall, London, 1931.

Various authors (R. S. Kirby et alii), *Engineering in History*. McGraw Hill, New York, 1956.

CHAPTER V

Towards the Industrial Revolution, 1500—1750

The sixteenth and seventeenth centuries mark the birth of engineering science. Until then, engineering had been an art based on empirical rules handed down from one generation to another by word of mouth and derived very much from trial and error experiments.

The scientific foundation on which the new engineering was built was laid down in this period by scientists such as Galileo, who showed that bodies of different weights fall the same distance under gravity in the same time, and who observed that the simple pendulum of given length has a nearly constant period of oscillation; by Newton, who formulated the laws of motion which have served engineers so well during the succeeding centuries and who invented the infinitesimal calculus (also invented independently by Leibnitz in Germany); by Napier, whose work on logarithms has eased the calculations of engineers ever since; by Robert Boyle, whose studies of the behaviour of gases under pressure explained the first steps taken in the heat engine; and by Robert Hooke the scientist, who pioneered so much in elasticity, the properties of materials and the design of instruments, clocks and watches. There were many others.

This was the era of the pump. Pumping water, and the ways in which it could be done, occupied the minds of both the laboratory scientists and the practical mechanics. This can be seen from the publications of Agricola, 1556,[1] and Ramelli, 1588—the latter describes more than one hundred different types of pumps—and all this activity gave birth in 1712 to that triumph of mechanical engineering, the Newcomen engine—which was not only a steam engine but a successful self-acting pump that did not depend on wind or water for its operation.

Rewards for inventors were encouraged by the establishment of letters patent (or patents) which were started in Venice in 1474 and in England in 1561. The Royal Society of London was set up in 1660 and started publication of its *Philosophical Transactions* in 1665. The movement to found learned societies where scientific matters could be discussed had begun a hundred years before in Naples, from where it spread to Rome in 1603 and later to Florence. The French Academy of Sciences started about the same time as the Royal Society of London and was followed by the founding of the Russian Academy in 1725 and the Berlin Academy in 1770. It appears that the need for scientists to get together and talk about their ideas and experiments away from the restrictions of religion was being felt in all the Western countries where scientific work was developing.

A number of the most brilliant scientists of the period were Englishmen and it has been suggested that they laid the foundations of English supremacy in engineering that endured throughout the seventeenth, eighteenth and nineteenth centuries. Another contributing factor was the extensive use of coal in England for all kinds of industrial purposes, and coal was the fuel used in abundance to operate the first heat engines of Savery, Newcomen and Watt. Only during the last few years has the primacy of coal as a fuel been challenged in the United Kingdom, though the situation has been different in many other countries for a long time.

One of the factors that helped to promote the industrial revolution in England was the excellent system of communication by water that had been established even before Brindley started building canals in 1760. Professor Skempton has pointed out that in the year 1600, 700 miles of naturally navigable river were available for transport, but that by 1760, 600 miles more had been added by building locks and weirs, deepening rivers and making cuts or canals, thus providing an important system of commercial transport. The cost of transport by river was about 3d. per ton mile compared with 1s. by road, and the loads that could be drawn by a single horse were very much greater, about 30 tons in a river barge as compared with $\frac{5}{8}$-ton on a soft road or $\frac{1}{8}$-ton by packhorse. Also the roads were frequently impassable in winter while the rivers were not. A feature of this development was the use made of Leonardo's invention of mitred gates for locks, used by him on the locks of the Milan canal in 1497; they were first used in England on the Exeter Canal in 1564–7 and four years later on lock gates in the River Lea at Waltham Abbey, after which they became standard practice. By 1750 there were few towns in England without near access to this system of river transport and the tonnage of cargoes was very substantial. For example, the Thames was navigable up to Lechlade where the river could take a barge of 50 tons; Tyneside coal was taken by the River Ouse to York and to Bedford; Leeds and Sheffield had achieved navigation by water, and works on the River Mersey and the River Irwell had brought navigable water to Manchester. On the River Weaver in Cheshire the annual tonnage of cargo ten years after the river became navigable was 31,500 tons and by 1760 it had increased to 60,000 tons.

MATERIALS

Two striking changes in the use of materials during this period were the gradual replacement of wood by metal, particularly cast iron, for the construction of machines, and its replacement (and that of charcoal) by coal as a fuel, particularly for industrial processes such as baking, brewing, sugar refining and the making of bricks, soap, alum and glass. The change from wood to coal as a fuel was very noticeable in London, which took 33,000 tons of coal from Newcastle in 1564 but a century later took more than half a million tons, with the result that the smoke from coal-burning in London was a source of amazement and horror to foreigners who visited the city.

The abundance of cheap fuel in Britain was an encouragement to many industries to develop exports, for they were in a more fortunate position in this respect than most of their competitors abroad. The increasing demand for coal in turn encouraged the development of mining and the search for methods that would cheapen the processes involved in mining, especially the drainage of mines, which was a most costly business, since it required horses by the dozen to work the pumps, or involved diverting streams and rivers to supply the water wheels that were the only alternative source of power until the advent of the steam engine. To transport such a material as coal cheaply in large quantities from one place to another was a challenge to the mechanical ingenuity of the day. A simple solution was to take it by water, and wherever possible this was done, by coast-wise shipping or in barges on the rivers and canals that were extended for the purpose. Near the mines, haulage for short distances overland was inevitable, and it was to make this easier that the horse-drawn tramway was developed. First running on wooden rails, later on iron rails, special trucks were made that could be hooked together to form a train to be pulled by a single animal. By 1700 horse-drawn railways were in general use in all the harbours and coalfields in Britain.

Wherever it was practical inclined railways were used, in which a train of full trucks hooked together descended an incline drawing a train of empty trucks up the incline as they did so. Later, as the quantities to be transported became still greater, it was not by accident that in the coalfields of the North of England the problem was first solved and the steam locomotive and the railway took over from the earlier methods of transport.

The coal industry of Britain was greater than that of any other country during the sixteenth and seventeenth centuries and the premier position of the country in the industrial revolution that followed was based upon the abundance of cheap coal.

There were no spectacular advances in metallurgy but the process of making iron was gradually improved in this period, largely because of the spread of knowledge due to printing. Agricola's work, *De Re Metallica*, is an example. While larger blast furnaces and better slagging techniques led to improvements in iron making, as did the use of power-driven tilt hammers worked by water wheels, the great demand for iron led to the rapid denuding of forests wherever iron was made. This situation contributed to the supremacy of Sweden as an iron producer for a time, for the purity of her ores and the extent of her forests placed her in a particularly favourable position, until in the year 1709 coke was used successfully for smelting iron in Shropshire and the end of dependence on charcoal was in sight. When in this period the use of wood as a constructional material for machines began to give way to metal, water tanks and pipes were made of lead or pewter which were both much cheaper than iron. According to J. W. Hall* in 1591: 'A forge could not make more than two tons of iron per week and owing to water shortages could often make only 50 tons per year.' Even a century later wrought iron was made in small quantities of 50 to 100 lb. at a time.

One of the most wonderful castings ever made is the great bell of Moscow that was cast in 1735 to replace a bell of 143 tons that had been cast in 1651. It weighs about 193 tons, is 20 ft. 7 in. high and has a diameter of 22 ft. 8 in. (By comparison, the weight of Big Ben, which was cast in Whitechapel in 1858, is $13\frac{1}{2}$ tons.) It was so heavy that the moulders could not get it out of the moulding pit and it was finally hoisted into position in 1836 by the army. Soon after it was cracked during a fire in the Kremlin, so was never rung.

During this period the scientists began making discoveries of direct use to mechanical engineering. Naturally not all of their work

* *T.N.S.*, vol. viii, p. 40.

was immediately applied to the making of machinery, for in many cases machines were built by men who had no knowledge of what scientists were doing; however, it would be wrong to suppose that the scientists were working away in ivory castles and that they had no contact with the mechanical problems of their day. It was quite otherwise in numerous instances as we shall see. This was the time when applied mechanics became a popular subject of discussion, argument and experiment. Applied mechanics can be said to have been started by Galileo, who demonstrated in his famous experiments on falling bodies at Pisa, not only that all bodies fall the same distance in the same time, but also that the velocity attained in falling is proportional to the time, and the distance travelled proportional to the square of the time of fall. Isaac Newton, a century later, explained the distinction between the mass of a body—that property which provides its inertia—and its weight, the external force of gravity acting on the body. While the mass is always the same, the weight of a body varies according to its distance from the centre of the earth. Newton propounded his three general laws of motion that have been so useful to engineers in determining the behaviour of machines, namely that:

(1) Bodies at rest or moving uniformly in a straight line remain so until acted upon by some external force
(2) Force is the product of mass and acceleration
(3) To every action there is opposed an equal and opposite reaction.

These generalizations finally solved the problem which had puzzled scientists since Galileo's demonstration: why should bodies of different masses fall to the earth with the same acceleration? The same generalizations explained the movements of the heavenly bodies that had been observed by Kepler long before and showed that the concepts of force, mass and acceleration were of universal application.

Galileo also lectured and wrote about strength of materials and made some elementary analysis of the strength of beams and cantilevers. For a simply supported beam of length l with a point load acting at distance x from one support he found that the bending moment is greatest under the load and is proportional to the product $x(l-x)$ so that the beam is most likely to break when the load is at mid-span. His consideration of the strength of a cantilever beam of uniform cross section led him to appreciate that if the width exceeds the thickness the beam is stronger when on edge than if it is lying flat, and this in the ratio of the width to the thickness. He even derived correctly the parabolic form of a cantilever of equal strength throughout its length having a rectangular cross section. He discusses in his book *Two New Sciences* the strength of geometrically similar structures and concludes that the larger they become the weaker they are, due to their own weight.

A name known to all mechanical engineers is that of Robert Hooke, because the proportionality of the pull to the stretch in the tensile loading of elastic materials—the so-called Hooke's law—is the very foundation of elastic theory on which so much mechanical design depends. In 1662 he was appointed the first curator of the Royal Society and among the experiments he demonstrated was the stretching of a suspended wire by weights placed in a scale pan at its lower end. He measured the distance between the pan and the floor as extra weights were added and found that if the load was not too great, the stretch was directly proportional to the weight. Hooke also experimented with spiral springs and coil springs for watches and clocks.

Another contributor to research in strength of materials at this time was the Frenchman, Marriotte, whose name has been coupled with that of Robert Boyle as an independent discoverer of Boyle's law. Marriotte carried out experiments with beams and produced a theory of stress distribution in elastic beams that was more satisfactory than those of his predecessors, but there remained a discrepancy between his analysis and his experiments that he was unable to resolve. He made some interesting tests on the bursting strength of thin pipes under internal hydraulic pressure. By altering the thickness and diameter of the pipes he deduced from his results that the thickness should be proportional to the product of the internal pressure and the diameter of the pipe—the first appearance of the thin cylinder formula for determining the thickness of pipes under internal pressure. For obtaining the highest pressures Marriotte used a column of water nearly 100 ft. high. In the mechanics of solid bodies Marriotte made an important contribution by demonstrating the conservation of momentum using balls suspended by threads to strike one another by impact.

Towards the end of this period a start was made by the Bernoulli brothers on the study of the shape of the deflection curves of elastic beams using the powerful methods of the differential calculus.

The first steelyard type of machine for testing the strength of materials appeared in Holland at the University of Leyden where

Professor van Musschenbroek built a tensile machine (Fig. 82) that was capable of pulling specimens of metal about one-tenth of an inch diameter and of wood about twice the size. The testing machine for materials did not come into general use for another century.

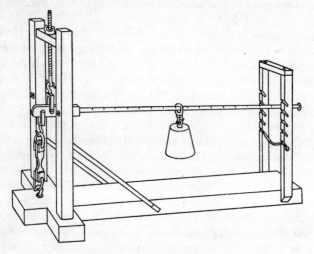

Fig. 82. Tensile machine of Musschenbroek.

TOOLS

The principal changes in the use of tools during this period arose on the one hand from the needs of the scientists for instruments of precision—for example, the microscope, the telescope and the clock—and on the other from the application of more power—from water wheels—in the working of metals. Rolls, both plain and grooved, were used for rolling metal, and large tup hammers for forging. There were also considerable advances made in the lathe, particularly by Besson. The appearance of rolling machinery for working iron marks a most important step forward in the application of power to the working of metals. It is believed to have originated in Sweden during the seventeenth century and to have been brought to England soon afterwards, primarily as part of the process of slitting wrought iron into rods cr bars for the hand-made nail trade and for other purposes. The metal was first heated in a furnace to a bright red heat and then successively rolled and cut in the slitting mill whilst it was still red hot. The operation can be seen in Fig. 83 and the arrangement for the drive of a similar mill in Fig. 84. Both of the rolls of each mill were separately driven by the two water wheels, the plain rolls being about 8-in. diameter and the slitting rolls about 12 in. Water was sprayed on the rolls to keep them cool. According to Jenkins there were in 1785 sixteen rolling and slitting mills in England, rolling 800 to 1,500 tons a year each, as well as some Irish mills which rolled about 2,000 tons a year between them.

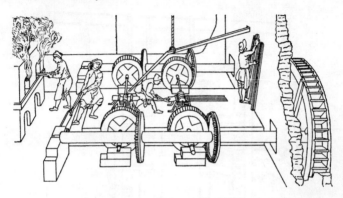

Fig. 83. Slitting mill, 1734.

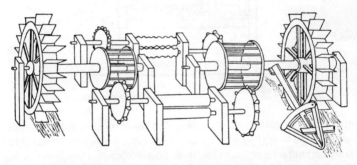

Fig. 84. Mill for iron-work, 1758.

Probably the most advanced machine tools of the period were those used for gear cutting—for making gear wheels for clocks and for scientific instruments. There is a wheel-cutting engine in the Science Museum dated 1672 that is certainly similar to one used by Robert Hooke and mentioned in his diary.[2] The formed cutter is driven through multiplying gears from a hand crank, while the screw at the front provides for adjusting the depth of the cut and the indexing plate for moving the work round from one tooth to the next. The machine itself is made entirely of metal and there are spikes, presumably to fasten it to a heavy wooden base. It should be noted that since the cutting wheel swings on pivots, the spaces between the teeth on the work would be left as concave depressions. By 1729 the Swedish engineer Christopher Polhem had introduced a hand-

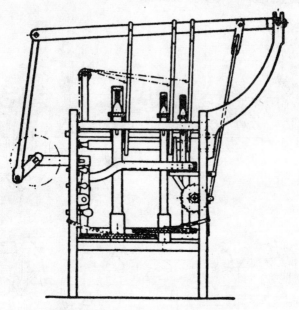

Fig. 85. Polhem's gear-cutting machine.

operated gear cutting machine (Fig. 85) using reciprocating broaches —this was one of the earliest uses of broaching but soon afterwards the same manufacturer introduced a power-operated machine using rotary cutters. Still more advanced was Hindley's wheel-cutting engine (Fig. 86) seen by Smeaton in 1741. This had a hypoid gear drive for the indexing plate which had reamered holes to engage a tapered pin for differential indexing on the index plate—a feature that was far ahead of its time and was only adopted on the milling machine one hundred and fifty years later.

Besson's screw-cutting lathe, 1579, is illustrated in Fig. 87. It shows the workpiece on the left hand side being turned with a taper screw thread, presumably for ornament. The same machine could

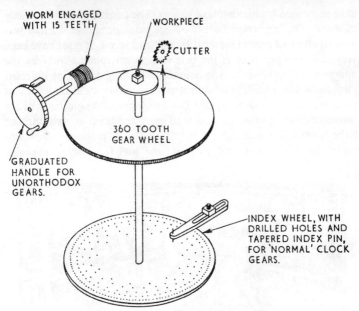

WORM ENGAGED WITH 15 TEETH

WORKPIECE

CUTTER

360 TOOTH GEAR WHEEL

GRADUATED HANDLE FOR UNORTHODOX GEARS.

INDEX WHEEL, WITH DRILLED HOLES AND TAPERED INDEX PIN, FOR 'NORMAL' CLOCK GEARS.

Fig. 86. Hindley's wheel-cutting engine.

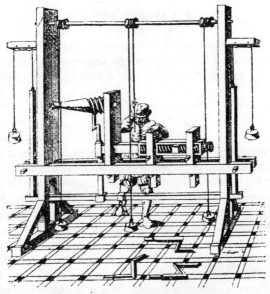

Fig. 87. Besson's screw-cutting lathe.

have been used for making lead screws. The construction was mainly of wood. The important feature of this machine is that the tool was held rigidly and moved forward by the lead screw. It must have been very slow in operation as the belt was pulled round slowly by the operator, nevertheless it was capable of cutting screws of different pitches by using pulleys of different sizes, either right or left hand with open or crossed belts, and Besson claimed it could be used for turning tapers or other contours. In another sketch an oval turning lathe is shown in which the tool was guided by a template containing a slot which was used to guide the tool for turning irregular

Fig. 88. Pole lathe.

shapes. Features of this arrangement were developed into the rose engine that was used for turning patterns for ornamental work which became very fashionable during the seventeenth and eighteenth centuries.

The lathe in most common use at this time was the pole lathe (Fig. 88) in which the work was directly driven by a rope around it, but subsequently, the mandrel lathe—anticipated by Leonardo—came into use during the sixteenth century (*see* Fig. 89). Here the work was driven by a live spindle or mandrel to which it was attached, the mandrel being supported at both ends and the work overhanging, if short, or otherwise supported at the end by a 'dead' centre. By 1700 iron had been substituted for the wood of the mandrel and soft metal collars or bushes let into the wooden headstock were used as bearings. The bushes of lead or tin were cast around the mandrel which had previously been machined between centres. At the end remote from the work the mandrel was threaded with short lengths of screw thread of different pitches. When the lathe was used for screw-cutting a guide plate was lowered to engage with the appropriate length of screw thread on the mandrel, whereupon the whole mandrel with the work would be moved axially as it rotated, thus producing a copy of the screw thread on the work, since the cutting tool was fixed to the bed of the machine (*see* Fig. 89). In the mandrel lathe the work both revolves and moves axially, the cutting tool

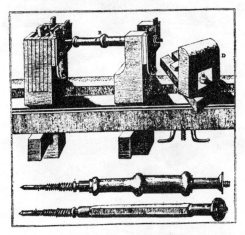

Fig. 89. Mandrel lathe.

being fixed, whereas in Besson's lathe the work revolved and the tool moved axially as in a modern screw-cutting lathe.

The use of screws for fastening is mentioned repeatedly in Agricola's description of mining machines, 1556[1]; for example, in describing a gear-driven chain pump he writes: 'The teeth of each wheel are fixed in with screws whose threads are screwed into threads in the wheel, so that those teeth which are broken can be replaced by others; both the teeth and the rundles are of steel.' It would appear from this description that screw taps for tapping holes must have been in use then.* The use of wood screws which were pointed and had a slot in the head seems to have been well known in Agricola's time for he quotes screws as being superior to nails for fastening the hide to the timber in making large bellows. A wood screw is illustrated at the bottom of his drawing (Fig. 90). Two other applications of the screw at this time are worth mentioning. Ramelli, 1588,

---

* Taps and dies for fine threads had been in use since the fourteenth century.

155

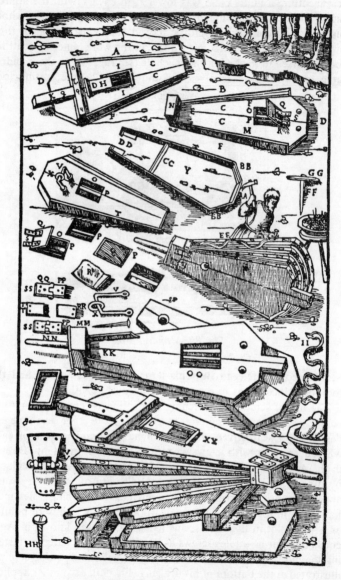

Fig. 90. Making bellows.

shows an illustration (Fig. 91) of a jack operated by a screw with left- and right-hand threads, and for making coinage in France the fly press with a balanced lever arm (Fig. 92) was introduced during the

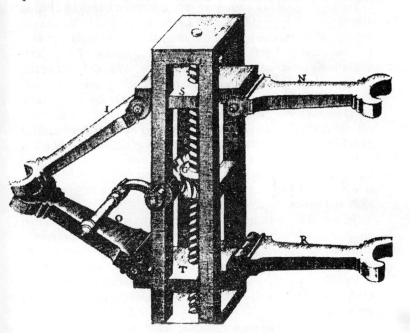

Fig. 91. Screw-jack of Ramelli.

Fig. 92. Balance coining press.

seventeenth century. This was an elegant way of storing human energy which could be released with little wastage provided the screw and its nut were well made.

The grinding machine was not a practical reality during this period although at its beginning in 1500 Leonardo had drawn sketches of disc grinding machines, and of internal and external grinding machines and polishers. There is no evidence that they were built until the beginning of the nineteenth century. What did exist at this time were large grinding wheels driven by power, of which an excellent example is shown by Woodbury[3] which also shows the water wheel for supplying the power. These grinding wheels were of naturally occurring sandstone and some were very large. Water was used both for cooling and to lessen the formation of dust but the grinding of hard steel with these stones was a very slow process. They were used for little more than sharpening and polishing. Smaller grindstones worked by a treadle were also in common use for sharpening tools and knives.

During the seventeenth and eighteenth centuries specialized machines were developed for the grinding of lenses for optical instruments and for the grinding of jewels. These were really lapping processes rather than grinding, for the cutting agent was in the form of a paste or powder of emery (a mixture of aluminium oxide, iron oxide and silica) or diamond dust, which was used for polishing jewels.

The instrument makers were responsible for introducing the micrometer, the vernier caliper and the vernier protractor during the seventeenth century, though their use in general engineering was not common until one or even two centuries later. The micrometer was invented and developed by astronomers for sighting telescopes and determining the distances between fixed stars. The originator, William Gascoigne, was killed at the battle of Marston Moor in 1644, but a description of his instrument was prepared by Robert Hooke and published with illustrations in the second volume of the *Transactions of the Royal Society* in 1667. The nature of the instrument can be clearly seen in Fig. 93. The measuring screw had left- and right-hand threads and for taking fine measurements a disc (not shown) divided on its edge into 100 parts was fixed to the spindle.

The spirit level—originally a glass tube filled with wine—was first used in France in 1666, though water levels of different kinds had been used many centuries before by the Romans and the Egyptians.

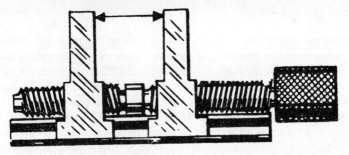

Fig. 93. Gascoigne's screw-micrometer.

Another useful instrument to make its appearance at this time was the spring balance. Its origin is unknown but the two illustrations (Fig. 94) published in 1694 show two forms—on the left an enclosed spiral spring with calibration marks for measuring force on the draw-bar and on the right a form with two pulleys connected by a cord, the upper pulley with a scale marked on its edge and containing a clock spring.[4]

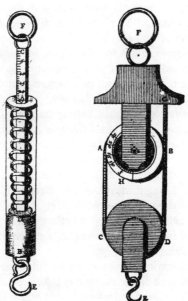

Fig. 94. Spring balance.

MACHINES

In the field of mechanisms the outstanding achievements of this period were the beam engine, the study and use of the pendulum, particularly in clockwork, the analysis of the forms of gear teeth, the application of the cycloidal form to actual gears before 1700, and the appearance of the wheelbarrow and the tramway in mining. It was not until the very end of this period that within the space of a few years a series of mechanical inventions were made in England that completely revolutionized the production of textiles. These were the flying shuttle, roller spinning, the spinning jenny, the mule and the power loom. Mechanization of what had been a cottage industry put the textile manufacture of England in a dominating position in the world for the next two centuries. These machines will be discussed in the next chapter.

The use of the pendulum in clockwork started a new era in mechanical timekeeping. Those mainly responsible for this important development were Galileo and Huygens. It is said that Galileo realized the possibilities of using the simple pendulum when watching the swinging of some suspended lamps in the cathedral at Pisa. He timed the oscillation by his pulse and found it to be constant. Later he ascertained by experiment that the time of oscillation of a pendulum was dependent upon its length, but independent of the amplitude of its swing. Many years later when he was blind he proposed to his son that the pendulum should be combined with a pin-wheel escapement in a mechanical clock, and the resulting model proved much more accurate than any mechanical clock that had been made before, showing variations of only a few seconds per day.[5]

Huygens, whose work on clocks began in 1656 and continued for twenty years, combined the pendulum and the verge escapement. He eventually concluded that for a pendulum to have an exactly constant time of swing it must move in the arc of a cycloid, and to achieve this he introduced cycloidal cheeks below the point of suspension of the pendulum so that its effective length was shortened as the amplitude increased. This was an unnecessary refinement that was soon dropped.

In 1670, after having served mankind for more than three centuries, the verge and foliot escapement gave place to the anchor escapement for mechanical clockwork (Fig. 95). This represented an advance almost as important as the pendulum itself. Unlike the verge it was able to operate with a very small amplitude of swing and being constructed with the faces of the pallets in the same plane as the teeth of the escape wheel, smaller clearances were required with consequently less recoil and less loss in friction. More important still, the pendulum was for part of its swing entirely free from the escape wheel, and thus the anchor escapement was the first step towards the ideal of a mechanical clock with an entirely free pendulum. The

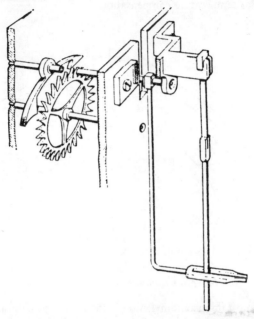

Fig. 95. Anchor escapement with pendulum.

standard length that was most commonly used was 39 in., resulting in a beat of 1 second, but in 1676 two clocks were built for Greenwich Observatory with 13-ft. pendulums having a beat of 2 seconds. A significant improvement to the anchor escapement was made in 1715 when Graham introduced his 'dead beat' escapement (Fig. 96). It was so called because the pendulum could swing—or beat—without producing any recoil in the escape wheel or the train of wheels behind it. In the original anchor escapement the pallet strikes the tooth of the escape wheel as it enters, causing a recoil, whereas in the improved 'dead beat' version, the face of the pallet slides down the face of the tooth of the escape wheel without causing any recoil. The amplitude of the swing of the pendulum is also less in the dead

F   161

beat escapement and this again leads to greater accuracy of time-keeping.

The effects of changes of temperature on the length of the pendulum and thus upon the accuracy of timekeeping were appreciated during this period and a satisfactory solution had been found before 1750 in pendulums filled with mercury and in the use of bi-metallic strips for temperature compensation.

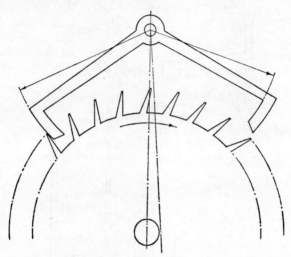

Fig. 96. Graham's 'dead beat' escapement.

The most important invention in the pocket watch was that of the balance spring—or hair spring. Unfortunately, there was some controversy for a time as to who had invented it, particularly as between Robert Hooke in England and Huygens in Holland. There is now no doubt that Hooke was the first inventor of the balance spring and he devised it to make a spring-driven movement sufficiently accurate for determining longitude at sea. He was unsuccessful in his attempts to form a syndicate to exploit his invention, so he never constructed such a chronometer. The story of how this was done and of the large sums of prize money offered by Governments for success has been well told by Usher.[6] One of the critical problems involved was temperature compensation of the balance wheel, finally achieved by using laminated bi-metallic strip, earlier attempts to compensate by altering the length of the balance spring having been unsuccessful.

The escapement generally used for watches was the verge and

balance spring, superseded in 1721 by Graham's cylinder escapement and balance spring. This was a dead beat escapement without recoil but it suffered from the defect that the escape wheel was never free of the balance wheel, and there was friction throughout the swing between the teeth and the cylinder. These defects were overcome in the lever escapement which came into use early in the nineteenth century.

It was soon evident to clockmakers that the shape of the teeth on the gear wheels should be such as to provide continuous contact with the minimum amount of friction. The mathematicians of the sixteenth and seventeenth centuries showed that the cycloid was the curve form which met these requirements. Later it was shown that the involute curve was even better.*

The properties of cycloid curves were studied by several mathematicians in Europe during the seventeenth century and one of them designed and constructed near Paris some machinery with gears having epicycloidal teeth in 1660. Before 1750 two other Frenchmen, first de la Hire and later Camus, established the basic geometrical principles of the design of cycloidal gear teeth. The first, who applied his principles to the design of gearing for a large water works, laid down that 'the aim should be to secure uniform pressure between the teeth and uniform motion; that tooth surfaces should be designed to roll on each other and so avoid friction; and that if a tooth of a gear is formed by a part of an epicycloid the tooth of the follower should be part of a hypocycloid described by the same generating circle'. Camus realized that even with the cycloidal tooth form there must be some sliding between the teeth with resulting friction and wear. He considered bevel gears and used the principle of the rolling cone for their analysis. He also analysed the combinations of a spur and a lantern gear and a crown gear and a bevelled lantern. Some of the teeth of the pinions he designed were heavily undercut and showed a great deal of back lash (Fig. 97). The developments of this period have recently been well reviewed by Woodbury.[2]

It must not be supposed that accurately cut gear teeth were in general use throughout this period or even by 1750. For larger machinery very crude cog wheels or even pins and lanterns were the order of the day, for example, the foot mill shown by Agricola,

* The cycloid is the curved path traced out by a point on the circumference of a circle as it rolls along a straight line. If the circle rolls round the *outside* of a larger circle the curve generated is called an epicycloid, and if it rolls round the *inside* of a large circle it is called a hypocycloid. An involute is the curve traced out by the end of a tight cord as it is unwrapped from a base circle.

1561 (Fig. 98). This picture and others in Agricola's book *De Re Metallica* are among the first in Europe to depict the wheelbarrow, one of the simplest and most useful machines ever invented. It appears that the rotating platform type of footmill or treadmill was in use in mines in Germany at that time.[4] A similar machine with an inclined

Fig. 97. Camus's epicycloidal gear teeth.

platform is shown by Ramelli, 1588 (Fig. 99). Jenkins asserts that in the middle of the eighteenth century cattle mills on this principle were still in use in Hungary.

When more power was required, and when neither water nor wind was available, then the ancient cage wheel or drum was used. The cage wheel, constructed of wood, was generally about 16 ft. in diameter and about 8 ft. wide, and the six or eight men treading inside it were able to lift 1 ton an average of 27 ft. forty times an hour, each man exerting about one-tenth of a horse-power. Cage wheels were in common use for unloading ships in the docks until they were replaced by steam and hydraulic cranes, though they became unpopular as soon as the treadmill was introduced into prisons (1818).

Fig. 98. Agricola: foot-mill, 1561.

A hoisting machine of particular interest is shown in Branca's book *Le Machine*, 1629 (Fig. 100). This shows two toothed wheels sliding on the same shaft, so that either can be engaged with the bars of a lantern wheel, the first illustration of a dog clutch for a reversing gear. One shudders to think how long it would last or what happened when there was delay in gear changing! Fig. 101 shows a dough-mixing machine, operated by cranks, from the same book.

Another hoisting machine of the time is shown in Fig. 102. This is a sixteenth-century pile-driver illustrated in Lorini's *Della Fortificazione*, 1596. Constructed mainly of wood and operated by hand cranks at each end of the main spindle, we see here an example of the use of the flywheel (the flywheel was being used increasingly in machinery), two pulleys to provide a mechanical advantage of 4, and a tripping mechanism, not unlike what would be used today.

Fig. 99. Ramelli: inclined foot-mill, 1588.

For handling materials Ramelli, 1588, illustrates a continuous bucket elevator operated by a hand crank, and Haselbery's book on mining, 1530, shows the first picture of a railway—or tramway—ever drawn. The rails were probably of timber. Horse-drawn tramways with timber and later metal rails were used extensively in

Fig. 100. Branca: reversible hoisting machine.

England from 1600 onwards. In the Newcastle district they were used to carry coal in wagons from the pits down to the riverside whence the coal travelled by ship to London and elsewhere.

That most elegant device for power transmission known as the Hooke's joint was invented by Robert Hooke and described in his

Cutlerian lectures in 1674. In its simplest form it is shown in Fig. 103 and is still used for connecting two rotating shafts whose axes meet in a point, so that the rotation of one shaft communicates rotation to the other, though with varying angular velocity. Another form of the same idea is embodied in the double Hooke's joint (Fig. 103) which is still in use for connecting shafts which may be either inclined or have parallel axes.

Fig. 101. Branca: dough-kneading machine, 1629.

In 1724 Leupold produced a textbook *Theatrium Machinarium* which has been described as the first attempt to produce a systematic treatise on mechanism. It has chapters on cams, on the crank, and on machines for converting circular motion into rectilinear motion. Although it was concerned with the strength of members as well as their motion, its importance lies in the fact that it was one of the first attempts to consider the constituent elements of a whole machine separately and examine their motions and constraints.

A new type of machine—the calculating machine—made its appearance during this period, two of them being of particular interest to mechanical engineers. The first was a device known as 'Napier's rods' or 'Napier's bones', comprising a set of uniform rods marked off with diagonals and numbers, the numbers in the triangular spaces formed being relevant products from the multiplication tables so that by placing these rods together in suitable ways the

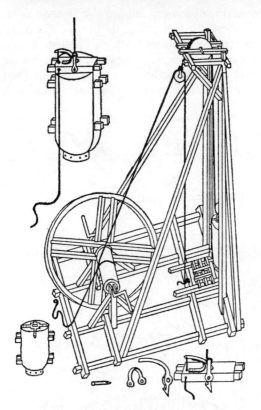

Fig. 102. Sixteenth-century pile-driver.

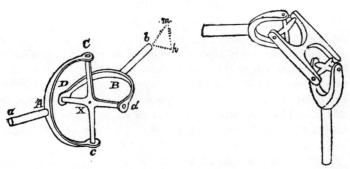

Fig. 103. Hooke's joints.

operations of multiplication and division could be performed. Napier's rods were the ancestors of the present slide rule. Secondly, the first mechanical computing machine, Fig. 104, was invented by Pascal in the seventeenth century. He constructed about fifty of these machines and obtained patent rights preventing anyone else from building computing machines of any kind during the life of his patent.

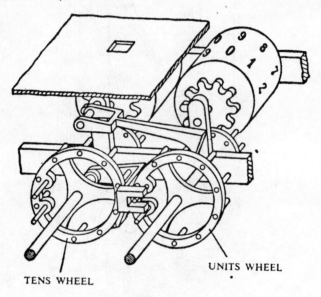

UNITS WHEEL

TENS WHEEL

Fig. 104. Pascal's calculating machine.

FLUID MACHINES

It was stated at the beginning of the chapter that this period saw the beginning of engineering science and that it was the era of the pump. It is commonly supposed that the stimulus for developing the pump arose from the need to drain water from the mines, but we shall now see that scientists in Europe, merely out of curiosity, had made tremendous strides in the science of fluid mechanics, so that it had become very much easier for engineers to build satisfactory pumps at the end of this period than it had been at the beginning.

Their outstanding achievement was the discovery of the pressure of the atmosphere and the consequent disposal of the old bogey that 'Nature abhorred a vacuum'. Torricelli's invention of the mercury barometer in 1644 was followed by Pascal's suggestion three years later that readings of it should be taken at the top and bottom of a mountain four thousand feet high and when this was done it disposed of any doubts that atmospheric pressure really existed and varied from place to place and from day to day. Another conclusive experiment made by Pascal was to construct one barometer inside another to show that the inner one, having no external pressure, was unable to support a column of mercury.

During this period laboratory experiments on fluids became more frequent, and their accuracy improved. Marriotte, for example, in 1686 made experiments on the impact of jets both with water and with air and among his conclusions was: 'Jets of water of unequal velocity will support by their impact weights which are proportional to the squares of their velocities'—the first appreciation that the force exerted by a stream of water is proportional to the square of the velocity of flow.[7] Like others he experimented with the trajectory of jets of liquid issuing from pipes and orifices using mercury as well as water. He was among the first to observe the resistance to flow in pipes and to deprecate sudden changes of direction of a pipe line because of the resulting increase in resistance. Before Marriotte, Torricelli had appreciated the principle of continuity of flow, and had realized that for a steadily flowing stream the velocities must be inversely proportional to the cross sections. He had also discovered that the velocity of water issuing from an orifice was proportional to the square root of the height of the water above the orifice.

The most important discovery of this time to engineers was that of Pascal who in 1647, at the early age of twenty-four, published the results of experiments he had made with syphons, syringes, bellows and tubes of different sizes and shapes, using different liquids. He concluded that 'in a fluid at rest the pressure is exerted equally in all directions', and illustrated this with his 'machine for multiplying forces' Fig. 105, anticipating the principle of Bramah's hydraulic press (see Fig. 165) by more than a century. Pascal's idea remained no more than a curiosity until Bramah converted it into a useful machine.

Some time later Sir Isaac Newton in his *Principia Mathematica* made two important contributions to fluid mechanics. The first concerned the notion of viscous shear in fluids. He showed analytically that if a cylinder rotates about its axis in a fluid at constant speed, a velocity gradient is set up in the fluid such that the velocity varies inversely with the radial distance from the axis. The other was the principle involved in the use of wind tunnels or water tunnels for studying the flow round objects immersed in a moving fluid. He asserted that the action of the medium on the body was the same whether the body moves through the medium with a given velocity or the medium moves at the same velocity past a stationary body. Newton also introduced the conception of a contraction coefficient for the flow of fluids through orifices.

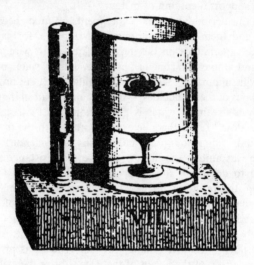

*Fig. 105. Pascal's machine for multiplying
forces.*

Among the experimenters of this time in fluid mechanics the best known is Henri Pitot who, in 1732, described what he called a 'machine' for determining the velocity of flow of a fluid. This was no more than two parallel small-bore tubes mounted in a frame, one straight and the other with a short right-angle bend at its extremity. The tubes were provided with taps and a scale so that after immersing the end of the frame in the flowing stream and closing the taps, the velocity of flows could be computed from the difference in height of the water in the two columns when the frame had been withdrawn.

Another experimental device that appeared before 1750 was the whirling arm for measuring the resistance of a body to movement through a fluid. The first apparatus of this kind, due to Robins

(Fig. 106), had a radius arm 4 ft. long that was rotated by a falling weight with cord and pulley. Tests in air and water were made with and without the body attached to the end of the radius arm.

In 1738 the science of hydrodynamics was born when Daniel Bernoulli published his *Hydrodynamica*. This substantial textbook contained the essence of the now famous Bernoulli equation, for he made the important assumption that for a flowing fluid the sum of the potential energy and the kinetic energy must remain constant. This equation, which has since been modified, notably by Euler, to include all forms of energy change, has formed the basis of the study of fluid motion ever since.

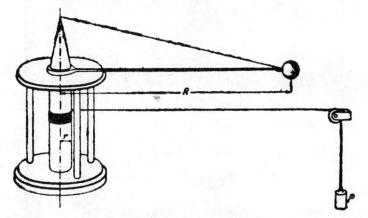

Fig. 106. Robins's rotating arm.

Details of piston pumps and other fluid machines as they were in use in the sixteenth century were recorded by Agricola in his famous book *De Re Metallica* published in 1556. This was the standard textbook on mining and metallurgy for over a century. The book is well illustrated and is remarkable for its accurate and detailed descriptions of machinery—both how it was made and how it was used. Agricola's favourite pump for mine drainage is illustrated in Fig. 107. Driven by an undershot water wheel 15 ft. in diameter, it worked three piston pumps simultaneously with a single crank of 1-ft. radius. The pump barrels were of wood. The pistons or buckets were of wood or iron without packing, perforated with holes and covered on the upper face with leather to act as a valve (*see* Fig. 108). Another type of bucket was a conical leather bag suspended point downwards

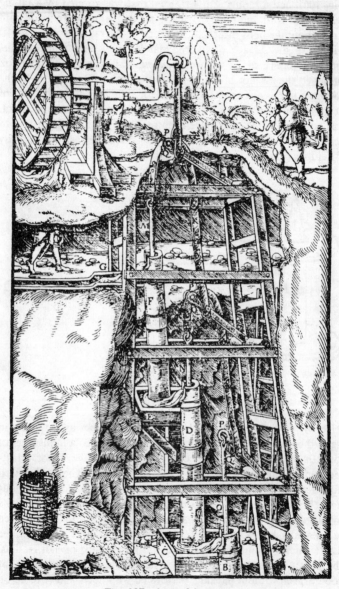

Fig. 107. Agricola's pump.

Fig. 108. Agricola: pump construction.

Fig. 109. Agricola. large crane driven by
overshot reversible water wheel.

from the pump rod. The bag collapsed on the downstroke and opened out against the cylinder walls on the upstroke.

Agricola also describes the chain of pots, and the rag and chain pump which had then been developed for lifts up to 200 ft. and could handle very large quantities of water. His largest machine was a reversible crane, worked by an overshot water wheel (Fig. 109) 36 ft. in diameter and provided with two sets of buckets in the circumference, one set being placed the opposite way to the other with suitable inlet sluices. The operator—Agricola calls him 'the director of the machine'—stands in a cage beside the reservoir and by drawing down the levers opens the sluices and controls the machine—the first illustration of a crane man in his cab. It has been estimated that this machine could develop 8 horse-power.

Ramelli, whose monumental volume on machinery appeared in 1588, described with 195 illustrations a large variety of machines including pumps, windmills, saw mills, screw jacks and derricks.[8] It is now generally accepted that some of the machinery he described could hardly have been made with the tools and equipment of his day. One hundred of his illustrations depict pumps and pumping machinery—only the most important can be discussed here. First we have a picture (Fig. 110) illustrating the principle of compounding, that is, there are two water wheels in tandem, the water passing first over the vertical overshot wheel and then on to the blades of the horizontal wheel. Note also that the two water wheels are both supplying power *to* the same system and that two distinct types of machine are taking power *from* the system; i.e. the grinding mill in the top left of the picture and the reciprocating saws cutting stone blocks at the bottom right of the picture. Secondly, in Fig. 111, we see a horizontal wheel which differs from the old Norse mill wheel in two important respects; the blades of the wheel are wholly above the water level and are struck by the incoming water only while they pass the inlet chute, so it is really a true impact wheel*; thirdly in the same picture the blades are cup-shaped, clearly a form superior to the straight blade for a horizontal wheel. Twenty years before Ramelli's work was published, Besson had shown a sketch of a semi-reaction wheel (Fig. 112) which came to be known as the tub or pit wheel. It was being widely used, particularly in France, by the beginning of the eighteenth century. Although the casing did not surround the whole of the rotor, one can see the outlines of the water turbine

---

* A horizontal impact wheel had earlier been described by Leonardo.

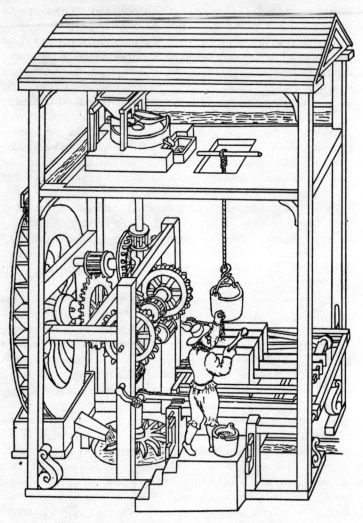

Fig. 110. Ramelli: compounding water wheels.

beginning to emerge and the blades were sufficiently curved for some reaction effect to have been obtained.

The final step was taken by about 1700 when a flour mill was built at Toulouse with the millstone direct-coupled to a horizontal water wheel that was completely enclosed in a casing. The vanes were beautifully curved throughout their length but the machine could hardly be described as a water turbine because there was such a large clearance between runner and casing that a substantial amount of water escaped to the tailrace without touching the wheel.

Among the many illustrations of pumps in Ramelli's work, there were a number of pictures of rotary pumps, of which the vane type

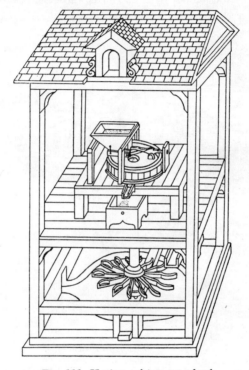

Fig. 111. Horizontal impact wheel.

shown in Fig. 113 was the most significant. This is what we now call a positive rotary pump. Its construction is shown in the bottom right-hand corner of Fig. 113, where the inner rotating member, **Z**, is seen to be eccentrically mounted within the casing E, so that the sliding vanes, which are loose, are moved radially outwards in their slots by centrifugal force. The extremities of the vanes are thus maintained in contact with the inner surface of the casing. The first true centrifugal pump was made, used and described by Dennis

Papin in 1689. Though Leonardo in 1500 had sketched a device for raising water by centrifugal force, the idea was not followed up, nor did Leonardo's sketch indicate an essential feature of the centrifugal pump, namely, entry of the fluid near the axis of the rotor and discharge at the circumference, both of which can be seen in Papin's so-called Hessian pump (Fig. 114). The early history of the centrifugal pump has been examined by L. E. Harris who describes its essential

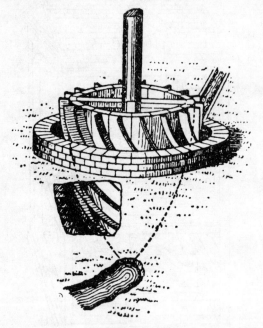

Fig. 112. Besson's tub-wheel.

feature: 'the rotation of the impeller produces a forced vortex in the water contained within the casing resulting in an increase of pressure in an outward radial direction and therefore a tendency to outward flow'.[9] Papin's machine meets these requirements though it had only two moving blades. He built others with improvements and he stated that his contrivance could be used for wind or water. It was certainly used for draining land near Marburg in Prussia. The centrifugal pump was not used for another century although during that period several patents were granted for centrifugal pumps, and in 1751 the mathematician Euler gave a mathematical analysis of its operation.

Fig. 113. Positive rotary pump.

Some of the developments in hydraulic machinery during this period can be seen by reviewing the changes in the London Bridge Waterworks between the time when they were established in 1582 and their final demolition in 1822. The first force pumps worked by a water wheel under one of the arches of the bridge were able to throw a jet of water over the steeple of St. Magnus Church—a feat that had never been seen before[4] and which gave confidence that the new water system would be effective in firefighting. The arrangement of the pumps in 1635 is shown in Fig. 115. A feature of particular interest was that the water wheel was a tide wheel which turned one way when the tide was flowing and in the reverse direction when the tide was ebbing. In both cases the crank and connecting rod oscil-

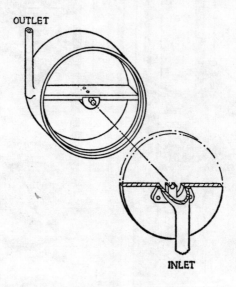

Fig. 114. Papin's Hessian centrifugal pump.

lated the disc to which the rods of the force pumps were attached. The delivery was into a large wooden pipe and thence to smaller pipes of lead. These works were destroyed by fire down to water level and by 1700 had been replaced by much larger works designed by Sorocold—a famous English engineer of the time. The new installation eventually comprised fifty-two pumps worked by four water wheels, one of which is shown in Fig. 116. Later, in 1761, another installation was added, designed by Smeaton and consisting of a water wheel 32 ft. in diameter by 15 ft. 6 in. wide, driving six pumps of 10-in. bore by 4 ft. 6 in. stroke.[10]

Sorocold designed many other successful water wheels and water-

supply works—for example, at Derby, Leeds, Westminster and Norwich—all to supply piped water. The wheel and pumps he installed
at Bridgnorth in 1706 were so successful that they were still working
in 1857. Sorocold suffered a curious accident in 1721 when showing
visitors a silk mill he had designed near Derby, for 'he slipped and
fell into the river and was carried under the 23-ft. dia. water wheel

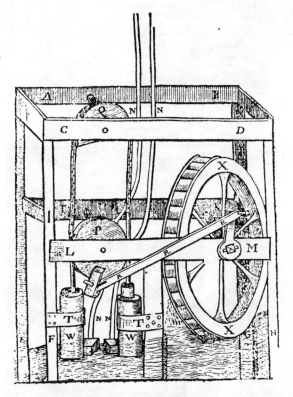

Fig. 115. London Bridge Waterworks, 1635.

which like Jonah's whale spewed him out to the mill tail where he
was taken up but received no hurt at all'.

It was then the usual practice for piston pumps to be worked by
water wheels so that even after the introduction of the Newcomen
engine for pumping at the mines we see in the 1733 viewbook of Amos
Barnes, a mine manager at Heaton Colliery near Newcastle, a sketch

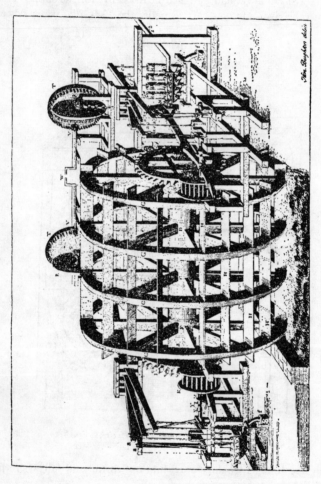

Fig. 116. London Bridge Waterworks, 1731.

of a water wheel operating at the bottom of a coal mine on water supplied by pumps worked by a Newcomen engine (Fig. 117).[11]

The most elaborate machine constructed for raising water was the famous one at Marly-la-Machine on the Seine, near Paris, for operating the fountains at Versailles. It was reputed to have been able to

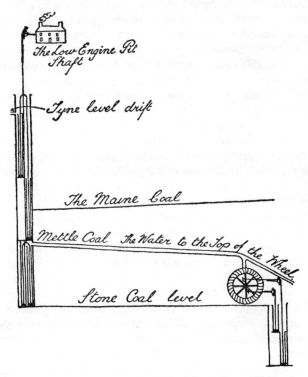

The Low Engine Pit Shaft

Tyne level drift

The Maine Coal

Mettle Coal   The Water to the Top of the Wheel

Stone Coal level

Fig. 117. Water wheel in Heaton coal pit, 1733.

perform a hundred horse-power of useful work but it was so in-efficient that the power actually produced must have been much larger. Built in 1682 at a cost of about a quarter of a million pounds, it was described within a few years as a 'monument of ignorance'. The machine was required to pump river water to a reservoir at the top of a hill 533 ft. high and three-quarters of a mile away. To do this 14 large undershot water wheels were erected in the river to drive in all about 250 reciprocating pumps. One-third of these pumps drew water from the river and pumped it to a cistern approximately one-third of the way up the hill and the water was then pumped the remaining distance in two stages with another cistern half way between, the 160 pumps at the two cisterns being driven by jointed iron rods reciprocated by cranks on the water wheels in the river 300 ft. below the upper cistern and half a mile away. The noise made by these rods in creaking and clanking when in motion was described at the time as sufficient to convince the most ignorant that the energy being expended was far beyond what was required for the purpose. The cost of maintenance was so great that when the water wheels became unusable about a century later, they were not renewed and the pumping was done by a steam engine of 64 horse-power.

The hydraulic air compressor was invented in Italy during the sixteenth century and was used widely on the Continent by iron-smelters for blowing forges without the use of bellows. The apparatus is in essential a water-operated aspirator or reversed air lift pump in which water from an overhead channel enters a vertical down pipe, at the top of which there are openings where atmospheric air can be entrained and drawn down to the bottom of the pipe by the velocity of the water. Thence it enters a submerged separating chamber from which the air is expelled through a pipe to the tuyère of the forge. Known as the 'trompe' or 'trombe', this device was used both for brass and iron smelting and possessed the great merit that no man-power or moving machinery was needed to supply the air blast. Fig. 118 shows an illustration of a trompe in use to blow a forge fire in the seventeenth century. In recent times the hydraulic compressor has been used to compress air up to 120 lb./sq. in. with efficiencies up to 85 per cent. Since large quantities of water have to be used, the air is compressed almost isothermally, that is, at nearly constant temperature.

Throughout this period the blowing of forges and blast furnaces in England was done by leather bellows of the type illustrated by Agricola, but in 1759 Isaac Wilkinson, the first of a famous line of iron-masters, patented a blowing machine of large size having three cast-iron cylinders and operated by a three-throw crank to give a uniform blast. He used these in his own works and also supplied them to others. He wrote afterwards 'I prospered. I grew tired of my leathern bellows and determined to make iron ones. Everybody laughed at me. I did it, and applied the steam engine to blow them and they cried "who could have thought it".'[12] A two-cylinder iron bellows of this type was in use at Dudden Furnace in Cumberland from 1785 until 1867, the cylinders being 5-ft. 2-in. diameter and

3 ft. 4 in. deep, fitted with iron pistons packed with leather. This blower supplied 3,000 cubic feet per minute when operated by an undershot water wheel 27 ft. in diameter.[13]

For the ventilation of mines Agricola describes a variety of rotary fans worked by the hands or feet of men, by animals, by water wheels

Fig. 118. Trompe, 1678.

and even by windmills which he concludes are the least suitable drive, 'for when the sky is devoid of wind as it often is, the machine does not turn'. Fig. 119 shows the construction of typical fans. The speeds were so low that they were necessarily positive displacement rather than centrifugal, and moreover the inlet and outlet ports were both in the periphery of the casing. In some fans the blades were tipped with goose feathers, otherwise the construction was entirely of wood except the metal crank handle and the four radial arms whose purpose was to act as a flywheel, for Agricola wrote 'they weight the axle, and when turned make it prone to motion as it revolves'. Like others at this time he seems here to hesitate in accepting the principle (not yet formulated) of the conservation of energy, so far as flywheels are concerned.

Fig. 119. Agricola: fans in mines.

Windmills were used increasingly during this period, not only for milling flour but also for pumping water, sawing wood—with reciprocating saws—and for many other uses. The numbers of windmills were very large indeed; there must have been thousands in England alone, and everywhere they formed landmarks which must have made a great difference to the appearance of the landscape.[14]

A remarkable feature of a windmill is the size of the sails—usually between 55 ft. and 80 ft. in diameter. To arrange for anything of this size to rotate at the top of a tower and deliver even a small amount of mechanical power presents a number of mechanical problems that were satisfactorily solved by the millwrights of the day. As we can see from Ramelli's drawing of a tower mill (Fig. 120) the construction was almost entirely of wood and this practice continued until Smeaton introduced cast iron into windmill work in 1754. The exception was the main bearings, where hard brass or gunmetal was commonly used, though some of the oldest were of stone, sometimes

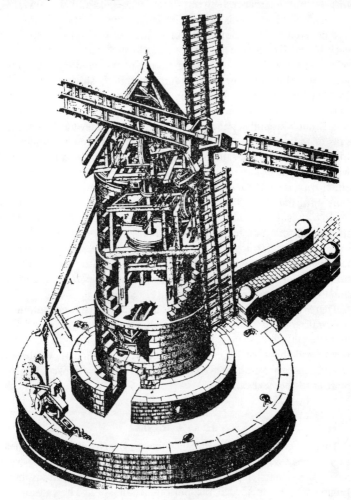

Fig. 120. Sixteenth-century windmill.

189

of marble, which became beautifully smooth and polished with use. Thick grease was used for lubrication; wooden worm wheels were cut by hand with a saw and finished with a chisel; bevel wheels were not used in windmills before 1750. For tower mills it became the usual practice to incline the axis of the journal carrying the sails about 5 to 15 degrees from the horizontal—partly to distribute some of the tremendous weight of the sails onto the tail bearing, and also to give more clearance between the moving sails and the tower. In the earliest mills the sails were pitched at an angle of about 20 degrees and covered with sailcloth laced in and out of bars on the frames of each blade. To stop the mill the cap was turned round until the plane of the sails was in line with the direction of the wind, and winching gear for doing this, that is, winding the mill, is shown in Fig. 120. In 1745 Lee patented a device for winding the mill automatically so as to turn it always to face the wind. Known as the fantail it consisted of a small windmill set at right angles to, and somewhat behind, the sails of the main windmill. While the wind was blowing square on to the main sails, the sails of the fantail presented no surface to the wind and did not move, but if the direction of the wind changed either way, the sails of the fantail began to rotate and turned the main sails until they were facing the wind again. The fantail was geared down very substantially; Wailes quotes an example of a mill where the fantail sails had to revolve six hundred times to turn the main-sails through 180 degrees and with a reasonable wind this would take nearly three minutes. This is one of the earliest examples of taking one of the controls of a machine from the operator and manipulating it automatically by mechanical means.[15]

There is little reliable information about the power output of windmills of this period. Those which operated two pairs of 4-ft. diameter grinding stones probably developed about 10 to 15 horse-power and this type were in the majority. According to Wailes the largest English windmill, which developed 50 horse-power, was not built until the next century.

HEAT ENGINES

There were two mechanical engineering achievements in heat engines that were in use before the end of this period. They were Savery's pulsometer pump, 1698, and the Newcomen Atmospheric Engine which superseded the pulsometer in 1712. The pulsometer pump, which came back into use for special applications and is still used for draining excavations and for handling water containing solids, may be defined as 'a steam pump in which steam is admitted alternately to a pair of chambers, forcing out water which has been sucked in by condensation of the steam after the previous stroke'. The pulsometer has no moving parts other than automatic ball or flap valves, and its historic importance lies mainly in the fact that it was the first steam-operated pump to be made and used in substantial numbers. At that time it was a dangerous device because it required steam under pressure to work it, and the design and operation of steam boilers was not understood, so that many explosions and accidents resulted. The Newcomen engine, on the other hand, was safe. The steam to work it could be obtained from a cook's cauldron at atmospheric pressure and whereas for mine drainage the pulsometer and its boiler had to be placed inside the mine, the Newcomen engine was erected on the surface at the pit head where the pump rods were attached to one end of an oscillating beam, the engine piston being attached to the other. The Newcomen engine was by far the more important of these two inventions for it was the first to use the energy of steam to work a piston within a cylinder and was therefore the forerunner of all the piston-type heat engines that succeeded it.

It could well be argued that this was the most important single step that has ever been taken in the history of mechanical engineering. However, it came about as the culmination of a series of advances that had been made during the preceding two centuries and it is most likely that if Newcomen had not built the first engine of this kind someone else would have done so very soon afterwards; indeed Dennis Papin very nearly did, for he was experimenting with the condensation of steam in a cylinder containing a piston, a few years earlier, but he was not a practical mechanic and he was defeated by mechanical difficulties. Nevertheless, though Papin did not succeed in building a practical engine, he was the first to use steam to move a piston in a cylinder. He is usually remembered for his invention of the lever type of safety valve, used soon after this on steam boilers, but originally devised for his 'digester' or pressure cooker. This lever-type safety valve was one of the first automatic controllers. In 1606 della Porta had published a treatise in which he explained that water could be forced out of a tank under pressure by applying steam to the surface of the water and drawn up from a lower level by condensing steam above the water surface, thus producing a suction. Ten years later de Caus described a fountain he had constructed to operate on this principle. However the real stimulus towards the construction of a steam engine came from the discovery of the pressure of the atmosphere. Many scientists were experimenting with gases during the next century and we find the hand-operated pneumatic pump for producing pressure or vacuum, the first complicated machine to be

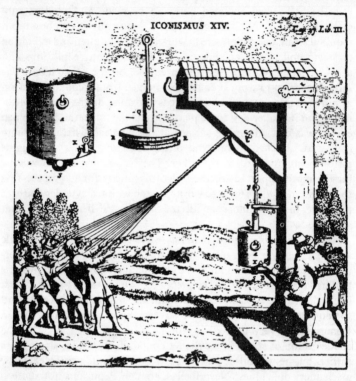

Fig. 121. Von Guericke's experiment with weights, 1654.

used in the laboratory of the scientist. These machines were made in large numbers before 1750 and they displayed an elegance of design that was seldom achieved in those days in full scale engineering.

Among the experimenters to use those machines were Boyle and Marriotte who before 1675 had announced the Boyle–Marriotte Law for gases—namely, that for a given mass of gas the volume varies inversely as the pressure when the temperature is constant.

A great public service to scientific men was performed in 1654 by the burgomaster of Magdeberg, Otto von Guericke, who demonstrated the force that could be exerted by atmospheric pressure in a series of dramatic but well-designed experiments that caught the public fancy and created a great impression on people thinking about the production of power. In one experiment sixteen horses in a tug-of-war could not separate two halves of an evacuated sphere; in another, heavy weights were lifted by a piston as a vacuum was created in the cylinder below the piston (Fig. 121).

In 1698 a master patent was granted to Captain Thomas Savery for 'an engine for raising water by the impellant force of fire'. There is reason to believe that this had been done thirty years before by the Marquis of Worcester with his 'water commanding engine', but it did not survive the experimental stage, whereas Savery's pulsometer engine known as 'the Miner's friend' was built in considerable numbers until it was superseded by Newcomen's atmospheric piston engine. The operation of Savery's pulsometer can be seen in Fig. 122. Steam was admitted alternately to each of the two egg-shaped vessels and starting with vessel P1 forced the water in it upwards through a check valve to the outlet pipe. When the first vessel, P1, was empty, steam was turned into the other vessel, P2, and meanwhile cold water was poured over the outside of P1 to condense the steam within and create a vacuum, thereby drawing water up from below to fill it. Meanwhile vessel P2 was being emptied and this sequence of events could continue indefinitely. These engines or pulsometers as we should now call them were made in a workshop in Salisbury Court, London, where notice was given in 1702 that they could be seen working 'on Wednesdays and Saturdays from 3 to 6 in the afternoons'. Although Savery's pulsometer achieved some success in pumping water for low lifts, there were many boiler accidents, mainly because the boiler pressures sometimes used—100 to 150 lb./sq. in.—were far beyond the safe limits of the crude construction of those days; for example, it was once reported that the heat was so great that the solder in the joints was melted and the pressure of the steam so high that several of the joints were blown open. Another complaint about the pump was that the amount of coal used was excessive. It has been estimated that the thermal efficiency of the engine, the ratio of the work done over the heat supplied (expressed as a percentage), was considerably less than one-half of 1 per cent.

The first reliable steam engine was the atmospheric beam engine erected by Thomas Newcomen in 1712 at Dudley Castle in Staffordshire (*see* Fig. 123). It was reported that 'the beam vibrates 12 times

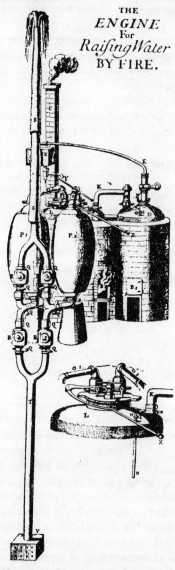

Fig. 122. Savery's engine.

in a minute and each stroke lifts 10 gallons of water 51 yards per-
pendicular'. This required the expenditure of nearly 6 horse-power
of energy in pumping or, as we should now say, the brake horse-
power was nearly six. Newcomen had the wisdom to separate the
pumping of the water altogether from the steam piston and cylinder,

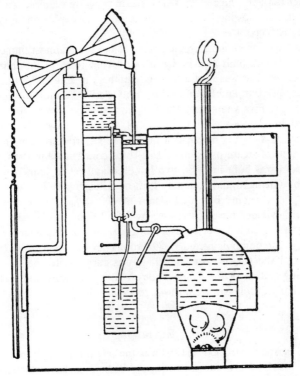

Fig. 123. Newcomen engine with manually operated valves.

and he used an ordinary plunger pump for pumping water, the
pump plunger being drawn up by the pressure of the atmosphere
pressing down on the top of the steam piston attached to the other
end of the rocker beam, while there was a vacuum below the piston.
The vacuum was brought about by injecting cold water into the
cylinder after it had been filled with steam. The steam piston was
raised again by the weight of the pump rods while the cylinder was
being filled with steam. The various events of the cycle—the admission of steam, admission of injection water and exhaust of condensed

and injection water were controlled by the engine itself, the valves being operated by rods attached to the 'walking or rocker beam'. To maintain a vacuum in the cylinder—which was too large for machining in those days—a leather sealing disc was used on top of the piston with a water seal above it, continuously replenished from a water tank at a higher level. The steam pressure required was little more than atmospheric, so the danger of boiler explosions was very greatly reduced.

The Newcomen atmospheric engine achieved an almost immediate success. It was built in cylinder sizes of from 10 to 72 in. diameter, and before 1750 was in use in large numbers in England, Hungary, France, Belgium and Spain and possibly elsewhere on the continent of Europe. Five years later the first atmospheric steam engine was erected in America. Beam engines of various kinds continued in production and it is perhaps of interest to record here that the last beam type of steam pumping engine to be constructed in the United Kingdom was built in 1919, two hundred and seven years after the first, by Glenfield and Kennedy Ltd. of Kilmarnock. It was not an atmospheric engine but used steam at 100 lb./sq. in. in the high pressure cylinder (26-in. diameter by 3-ft. 10½-in. stroke) the steam then passing to a lower pressure cylinder (40-in. diameter by 6-ft. stroke). The pump delivered 1,389 gallons per minute against 265-ft. head and the engine cost £8,500. It was supplied to the Eastbury pumping station of the Colne Valley Water Company and was still in use daily during the period 1942 to 1954.

REVIEW

The period from 1500 to 1750 was the one immediately preceding the Industrial Revolution, which may be said to have started with the general use of the steam engine. During this time both scientists and practical men were much engrossed in the problems presented by the pump, and we have seen how both contributed to their solution, but the invention of the Newcomen atmospheric steam engine in 1712 dwarfed every other mechanical achievement. Among scientists we have mentioned Galileo, Newton, Hooke, Huygens, Boyle, Marriotte, Pascal and Bernoulli, all of whom, among many others, contributed to the scientific explanation of the phenomena involved in making a successful self-acting pump. The men who contributed in their various ways to its practical achievement were Besson, Polheim, Wilkinson, Worcester, Von Guericke, Savery, Papin and Newcomen, but there were also many others who in small ways made useful advances in mechanical engineering during this period.

The other pumping devices that were born but not developed at this time were the positive rotary pump, the centrifugal pump and the trompe or water injector for blowing air. It is singular that the leathern bellows were superseded by iron bellows (by Wilkinson) without the Chinese wooden bellows having ever been used in England, though wooden box bellows had been made in Germany from 1550.

We can see that several steps had been taken towards the development of the water turbine that was to be evolved during the succeeding period—compound water wheels, the impact wheels, curved vanes on the rotor, the pit wheel, and finally the totally enclosed water wheel all being steps towards its evolution.

The use of long rods for driving pumps at a distance from the source of power, developed considerably after Agricola's time, particularly in southern Germany where *Stangenkunst* or 'pole-devices' were thought sufficiently important to be depicted on large coins and medals.[14] This type of drive reached its zenith in the famous waterworks at Marly (1681–8), which was the subject of so much comment and discussion that the inefficiency of such a kind of power transmission became apparent to all.

Apart from pumping, the outstanding mechanical achievement of the time was the pendulum clock, a development that also gave rise to improvements of lasting value to mechanical engineering such as the study of gearing, gear-cutting machines, broaching, and the screw-cutting lathe. Perhaps of special importance was the growth of a general realization of the significance of friction in machinery.

The subject that we now call strength of materials began to take shape during this period, though rather more slowly than one might have expected. Hooke's fundamental experiments in stretching a wire, which led him to the notion of elasticity, and Marriotte's experiments on the bursting of thin pipes under internal pressure, from which he deduced the thin-cylinder formula, were the first steps; but it was the application of the calculus to the analysis of the deflection of beams by the Bernoulli brothers at the end of the period that marks what might be termed the beginning of the analytical approach to the study of the strength of materials. At the same time these analytical methods were being supplemented by the use of the first tensile-testing machine for solid bars. Just as the Newcomen engine represented the most important step in making a machine, so the discovery and application of Hooke's Law (i.e. that stress is proportional to strain) was the greatest event in the field of strength of materials.

In the field of transport and handling materials, both the railway, or tramway, and the wheelbarrow were apparently developed in connection with mining. Hoisting machines were also improved as we have seen in the examples of the pile-driver and the bucket elevator.

All these were but trifles in comparison with the achievement represented by the Newcomen atmospheric steam pump. We may appropriately conclude this review by considering in more detail the nature of this achievement from the mechanical point of view. It was the first practical heat engine that could be used as a source of mechanical power; moreover it was the logical forerunner of all the piston-type engines that were to follow, since it comprised the essential elements of a reciprocating piston working in a cylinder, the piston being coupled to members so that it could perform mechanical work automatically. The piston was caused to move by periodic differences of pressure on its two sides, these pressure differences being brought about by the alternate application of heat and cold to the working substance in the cylinder. The working substance was steam, generated in an external boiler and admitted to the cylinder periodically through a valve worked automatically by the engine itself, while another valve, also automatically operated, admitted the cold injection water at appropriate intervals.

It is not easy for us at this distance in time to assess the magnitude of this achievement. Before the Newcomen engine there existed only three means by which mechanical power could be produced—by the muscular efforts of men or animals, by water wheels, or by windmills. Here in the steam engine was something entirely new, independent of the vagaries of men or of the weather; it would continue to do mechanical work, tirelessly, by night and by day, so long as fuel and water were fed to the boiler. Moreover, the power produced was not limited by the flow of the river or the force of the winds but could be made as large or as small as was suitable for the work to be undertaken. Finally, the automatic action of a sequence of events controlling its operation and following relentlessly one upon another gave a new conception of the possibilities of machinery, obviously greater than any that had arisen from the study of clockwork.

REFERENCES

(for abbreviations see List of Acknowledgments)

1. Agricola, G., *De Re Metallica*, trans. by H. C. and L. H. Hoover. Dover Publications, New York, 1950.
2. Woodbury, R. S., *History of the gear-cutting machine*. M.I.T.P., 1958.
3. Woodbury, R. S., *History of the grinding machine*. M.I.T.P., 1959.
4. Jenkins, R., *Collected Papers of Rhys Jenkins*. C.U.P. for Newc. Soc., London, 1936.
5. Singer, C. (editor), *A History of Technology*, vol. iii. O.U.P., 1957.
6. Usher, A. P., *A History of Mechanical Inventions*. Harvard U.P., 1954.
7. Rouse, H., and Ince, S., *History of Hydraulics*. Iowa Inst. Hyd. Res., State University, Iowa, 1957.
8. Parsons, W. B., *Engineers and Engineering in the Renaissance*. Williams and Wilkins, Baltimore, 1939.
9. Harris, L. E., *T.N.S.*, vol. xxviii, 1951–3, p. 187.
10. Beighton, H., 'Waterworks at London Bridge', *Phil. Trans.*, 1731, p. 442.
11. Raistrick, A., *T.N.S.*, vol. xvii, 1936–7, p. 131.
12. Davies, A. S., *T.N.S.*, vol. xxvii, 1949–51, p. 70.
13. Mart, J. N., *T.N.S.*, vol. xviii, 1937–8, p. 93.
14. Contributions from the Museum of History and Technology, *U.S. Nat. Mus. Bull.*, 218. Smithsonian Institution, Washington, D.C., 1959.
15. Wailes, R., *The English Windmill*. Routledge and Kegan Paul, London, 1954.

BIBLIOGRAPHY

Agricola, G., *De Re Metallica*, trans. by H. C. and L. H. Hoover. Dover Publications, New York, 1950.
Crombie, A. C., 'Augustine to Galileo', *The History of Science, A.D. 400–1650*. Falcon Press, London, 1952.
Dickinson, H. W., Papers in *T.N.S.*, vols. xx to xxvii.
Ewbank, T., *Hydraulic Machines*. Tilt and Bogue, London, 1841.

Finch, J. K., *Engineering and Western Civilization*. McGraw Hill, New York, 1951.

Hart, I. B., *The Great Engineers*. Methuen, London, 1928.

Jenkins, R., *Collected Papers of Rhys Jenkins*. C.U.P. for Newc. Soc., London, 1936.

Klemm, F., *A History of Western Technology*, trans. by D. Singer. Allen and Unwin, London, 1959.

Lilley, S., *Men, Machines and History*. Cobbett Press, London, 1948.

Matschoss, C., *Great Engineers*, trans. by H. S. Hatfield. George Bell, London, 1939.

Rouse, H., and Ince, S., *History of Hydraulics*. Iowa Inst. Hyd. Res., State University, Iowa, 1957.

Salzman, L. F., *English Industries of the Middle Ages*, 2nd edn. O.U.P., 1923.

Singer, C. (editor), *A History of Technology*, vol. iii. O.U.P., 1957.

Skempton, A. W., 'The Engineers of the English River Navigators, 1620–1760', *T.N.S.*, vol. xxix, 1953, p. 25.

Timoshenko, S. P., *History of Strength of Materials*. McGraw Hill, London, 1953.

Usher, A. P., *A History of Mechanical Inventions*. Harvard U.P., 1954.

Various authors (R. S. Kirby et alii), *Engineering in History*. McGraw Hill, New York, 1956.

Vowles, H. P. and M. W., *The Quest for Power*. Chapman and Hall, London, 1931.

Wolf, A., *A History of Science, Technology and Philosophy in the XVIth and XVIIth Centuries*, 2nd edn. Allen and Unwin, London, 1950.

CHAPTER VI

The Industrial Revolution, 1750–1850

During this short period of only one hundred years tremendous strides were made that transcended all that had been achieved before. In mechanical engineering the scene was dominated by the heat engine—in the practical sphere by enormous improvements to the steam engine and its application to all kinds of industrial purposes, to the railway and to shipping—and in the scientific sphere by the birth of the science of thermodynamics, for it was during this period that the properties of steam and its latent heat were discovered; that a unit for the rate of doing mechanical work (the horse-power) was introduced, that the mechanical equivalent of heat was established and thus the first law of thermodynamics; that the ideal cycle for a heat engine (the Carnot cycle) was enunciated, and lastly, that a clear statement was made by Clausius of the second law of thermodynamics stating that there is a limit to the efficiency of any engine which converts heat energy into useful work. The importance of these precepts was not immediately evident to many engineers, but they formed a solid foundation on which the elegant science of thermodynamics was built in the years that followed.

In practical matters Britain remained in the lead, as she demonstrated in the Great Exhibition of 1851 which has been described by Mumford[1] as 'the triumphant cock crow of the country that boasted itself the workshop of the world'. Here[2] were shown working exhibits of machine tools, steam engines, pumps, textile machinery and all the multifarious products of British industry. It was a great triumph for the engineers of the day, yet the men who made it were for the most part uneducated in the scientific knowledge of their time, and did not fully understand the principles of the operation of the machines they constructed.

It was not only the engineers who were behind the times, for in 1798 Dr. Thomas Young, who gave his name to the modulus of elasticity, remarked that Britain was very much behind her neighbours in many branches of mathematics.

An example of the disquiet felt by an eminent engineer, Sir William Fairbairn, appeared in his report of the Paris Exhibition of 1855 where he wrote:[3]

'The French and Germans are in advance of us in theoretical knowledge of the principles of the higher branches of industrial art; and I think this arises from the greater facilities afforded by the institutions of those countries for instruction in chemical and mechanical science. . . . Under the powerful stimulus of self-aggrandizement we have perseveringly advanced the quantity, whilst other nations, less favoured and less bountifully supplied, have been studying with much more care than ourselves the numerous uses to which the material may be applied and are in many cases in advance of us in quality.'

A century earlier students on the continent of Europe were being well trained in engineering science. Although the famous École Polytechnique in Paris was not officially founded until 1795 after the French Revolution, the École des Ponts et Chaussées for training civil engineers for government service had existed since 1747 and other engineering schools also required of their students a high standard in mathematics; for instance, it is recorded that Professor Bossut of the engineering school at Mezières in 1752 considered it his duty 'to teach young engineers the mechanics of fluids which is necessary to their profession'.[4] The training in engineering science given at the École Polytechnique set the standard for continental countries during the nineteenth century. It began with a two-year course in mathematics, mechanics, physics and chemistry and continued with courses by practising engineers about the design, construction and operation of machines and structures.

The German Polytechnics which were established soon after the Napoleonic wars followed the same plan but trained students also for industry and gave more time and emphasis to the study of machine design. By 1830 Polytechnics had been established at Berlin, Karlsruhe, Munich, Dresden, Stuttgart, Prague and Vienna. No such opportunities for engineering education existed in Britain. The first professorship of engineering was not established until 1840 (at Glasgow) and though two others followed soon afterwards at King's College and University College in London, it was not until 1875 that a professorship of engineering came into being at Cambridge. Though Britain was the leading industrial country of the world, the engineering problems that arose in the course of her development had to be solved by men who had taught themselves—most of whom had little understanding or appreciation of science.

An attempt to spread technical knowledge to the adult population was made in 1799 when the Royal Institution was founded in London,

'for diffusing the knowledge and facilitating the general and speedy introduction of new and useful mechanical inventions and improvements and also for teaching by regular courses of philosophical lectures and experiments, the applications of the new discoveries in science to the improvements of arts and manufactures and in facilitating the means of procuring the comforts and conveniences of life.'

Apparently this was not enough, so in 1831 a band of enthusiasts considered it necessary to found the British Association for the Advancement of Science—a body that flourishes to this day. Its avowed objects were:

'to give a stronger impulse and a more systematic direction to scientific inquiry, to give a greater degree of national attention to the objects of science, and a removal of those disadvantages which impede its progress, and to promote the intercourse of the cultivators of science with one another and with foreign philosophers.'

Later in its history the British Association established an engineering section, though this has not played a very prominent part in engineering affairs. The first meeting of the British Association took place in York, and one of its sponsors was the celebrated Yorkshireman, mathematician and scientist Sir George Cayley, who later founded the Royal Polytechnic Institute* in Regent Street London in 1839.[5, 6] He was chairman of its board for a number of years and is reported to have said: 'We much want a good scientific board, confined by no aristocracy of orthodox men who sit like an incubus on all rising talent that is not of their own shop. Freedom is the essence of improvement in science.' Most important to this history was the establishment in 1847 of the Institution of Mechanical Engineers with George Stephenson—the father of the railway—as its first president. It can be said that from that time the profession of

* Not to be confused with the founding of the Polytechnic, Regent Street, by Quintin Hogg in 1882.

mechanical engineering in Britain took on an air of professional respectability.

The advent of the steam engine gave a terrific impetus to all branches of engineering and indeed to every kind of industrial activity. Now at last the mines could be kept dry; more coal could be hauled more quickly to the surface; the steam locomotive was invented to move it from the pit head; more iron could be made with more air blast from a steam-driven blower; machine tools could be made bigger because there was a steam engine to drive them; bigger pumps could be made for the same reason, and perhaps more important than any of these was the freedom that the steam engine provided to establish factories and towns wherever it was most convenient to do so, not necessarily at the riverside. By the year 1835 there were in Lancashire and the West Riding of Yorkshire 1,369 steam engines and 866 water wheels. A large proportion of these prime movers were serving the textile industries, which by this time had become one of the largest exporters of manufactured goods in the country. The value of British textile exports rose from £46,000 in 1750 to £46,000,000 in 1850.

Another large export was coal. Coal in England was then cheap and plentiful. Means had been found for using coke in place of charcoal in the blast furnace for making iron; coal gas was used for domestic and street lighting; wrought iron was being produced by the puddling process and Huntsman had introduced his process for making crucible cast steel.

Transport and communications in England continued to improve; canals were being built; roads were being improved by Macadam and the turnpike system, and above all the railways rapidly covering the country towards the end of this period provided better and faster communication than had ever existed before—so did the penny post, established in 1840.

MATERIALS

During the whole of this period England led the world in the manufacture of iron. In 1750 she produced 20,000 tons of pig iron but by 1850 her production had increased more than a hundredfold to something like $2\frac{1}{2}$ million tons.

Several factors contributed to the fantastic success of the English ironmaster during these hundred years. The principal ones were: the discovery of how to smelt iron with coke or anthracite instead of charcoal; the steam-driven blowing engines which gave a much stronger blast than used to be obtained with leather bellows: the application of Cort's puddling process and the use of grooved rollers in the making of wrought iron; and the use of Neilson's hot blast stove for pre-heating the air blast to the blast furnace.

Just before this period began Huntsman, a Doncaster clockmaker, had invented his process for making crucible steel in an attempt to obtain a better steel for clock springs.[7] His method was to melt up bars of blister steel mixed with flux in closed clay crucibles in a furnace so that the furnace gases could not contaminate the steel, and

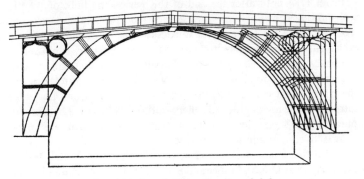

Fig. 124. Iron bridge, Coalbrookdale.

accordingly coke could be used for heating. He established works in Sheffield in 1742 and the process was successful for producing high quality steel for springs, tools, and cutlery, or indeed for anything where hardened steel was needed. But it was an expensive process, quite unsuited for the production of rails, heavy plates, heavy forgings or structural sections, and though its use continued for more than a century the annual production by 1850 was no more than 60,000 tons.

The use of cast iron for machine parts and structures began to grow rapidly after 1750. In England it was used successfully for casting cannon so that in 1773 it was reported that 'whereas, during the last twenty years not one English naval cannon had burst, in the French navy such accidents were so common that the sailors feared the guns they were serving more than those of the enemy'.

In 1760 the Carron iron works started casting large gear wheels, and in 1777 the first cast-iron bridge was built by Abraham Darby over the River Severn near Coalbrookdale.[8] The weight of cast iron used in the bridge was 378 tons. The many cast-iron parts were locked into one another and held fast by wedges without any bolts or rivets. The sequence of events for assembly and erection on the site was tried out on a wooden model (Fig. 124) for which a gold medal was awarded by the Society of Arts in 1788.

Other examples of the use of cast iron were in the all-metal machine tools that were being developed, the all-metal equipment of the famous Albion Flour Mills in London, designed by Rennie, water pipes for all purposes, the cylinders and pumps of steam engines and above all for domestic hollow-ware.

From 1784 until after the end of this period the bulk of the pig iron produced in England was used for the production of wrought iron. In that year Henry Cort introduced the puddling process in which the pig iron was melted down in a reverberatory furnace and the pasty mass stirred by the puddler with a long iron bar so as to cause enough air to come in contact with the hot metal to burn out the carbon and convert it into malleable wrought iron. Cort also introduced the use of grooved rollers—originally invented by Christopher Polhem[9]—so that the output from his equipment was much increased, resulting in lower production costs.

About this time it was established by the English scientist, Joseph Priestley, and others in France and Sweden that it was the combination of carbon with iron that produced steel.

There is special interest for the mechanical engineer in the invention by Neilson in 1828 of the hot blast stove for pre-heating the air to the blast furnace (Fig. 125). Its significance to the iron master was that when the air was pre-heated 600° F., three times as much iron could be produced for the same amount of fuel; iron ore which had been unusable with the cold blast could now be used; the temperature was so high that coal or anthracite could be used as fuel and many impurities previously retained were burnt out; indeed the high temperature made it necessary to watercool the air nozzle or tuyère and this was done by inserting a coil of wrought-iron pipe in the tip of the cast-iron nozzle. The Neilson air pre-heater was a cast-iron vessel containing air passages enclosed in brickwork and fired by solid fuel burning on a grate at first and after 1834 by the hot gases taken from the top of the blast furnace. The hot blast stove is historically of considerable importance to mechanical engineering since it is one of the first examples of a piece of heat transfer equipment being added to a process simply in order to raise the temperature and save heat. The cup and cone device for closing the top of the furnace—previously left open so that the gases burned to the heavens—was introduced in 1834. This allowed charging when required but saved fuel and made the furnace easier to operate.

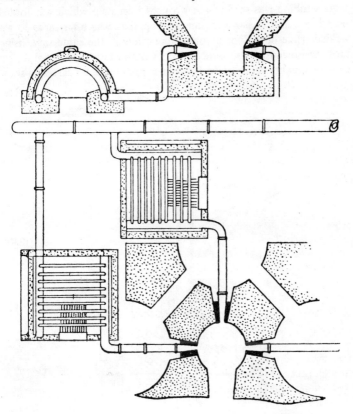

Fig. 125. Neilson's hot blast, 1829 and 1832.

Considerable improvements took place in forging, particularly after the invention of Nasmyth's steam hammer (Fig. 126, Frontispiece) in 1839. Much larger forgings could be worked under this hammer than was possible with the tilt hammers used previously; also, it was possible, after the steam cylinder had been made doubleacting, to vary the force of the blow from its maximum down to just sufficient to crack an egg-shell. Another use for the steam hammer was in pile-driving for civil engineering works. It was used for this purpose in 1849 for making the foundations for Robert Stephenson's high-level bridge at Newcastle upon Tyne. The hammer delivered one blow every second compared with one every few minutes by the methods used before.

Iron tubing was made from wrought iron strip which was folded round a mandrel bar with an overlap that was afterwards forge-welded while hot. Copper tubes were made in the same way, using brass to braze the joint until in 1838 Charles Green patented a process for making solid drawn brass or copper tubes by casting a short

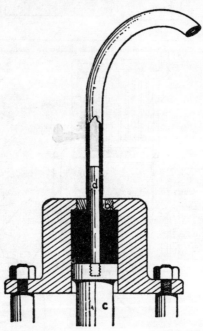

Fig. 127. Burr's extrusion press.

length in a split cast-iron mould containing a core and then drawing down through a die on a chain-driven draw-bench in much the same manner as is done today.

The squirting or extrusion of seamless lead tubes was started in 1820 by Thomas Burr, a plumber in Shrewsbury. He used Bramah's hydraulic press (*see* Fig. 165) to obtain the force required and squirted the metal through an orifice containing a mandrel (Fig. 127). A similar method was used later for sheathing electric cables with lead.

The technique of analytical chemistry became well developed at the beginning of this period; as a result a number of the less common metals were identified, and though these did not find immediate applications in engineering work, they were later of great value as the constituents of special steels. The dates when these metals were isolated were Nickel, 1751, Manganese, 1774, Molybdenum, 1781, Titanium, 1794, Chromium, 1797 and Vanadium, 1830.

Achievements during this period in strength of materials began with the publication in France of Euler's *Analytical Mechanics* which was the first textbook on mechanics in which the calculus was used to study the motion of a particle. Newton and others had used geometrical methods, but the analytical methods used in this book had a great influence on later developments in mechanics. Euler is, however, best known to engineers for the formulae he derived for expressing the critical load of columns. The critical load is that load when buckling begins under the strain of the applied load. Euler published a great many mathematical papers on strength of materials before his death in 1783 and for more than forty years afterwards the Russian Academy of Sciences continued publishing papers from his pen.

Some valuable contributions to the subject were made in 1807 by the English mathematician Thomas Young, whose name has been given to the modulus of elasticity, the ratio of the stress to the strain of a structural material, which is now called Young's modulus. Young showed that the modulus was equally important for tension and compression and that Hooke's Law applies only up to a certain limit—the limit of elasticity. Regarding a modulus for shear, Young pointed out that no direct tests had been made, but that it could be inferred from the properties of twisted circular shafts that the shear stresses are proportional to the distance from the axis and to the angle of twist. He also made a study of the stresses produced by impact and coined the term resilience to express the power of a body to resist very rapid motion.

Two French engineers, both graduates of the École Polytechnique, made experiments on beams which confirmed theoretical analysis and served to give designers confidence in the application of analytical methods to engineering problems. The first was Dupin, who published in 1815 the results of his tests on the bending of wooden beams. For a beam supported at both ends he compared the deflection produced by a concentrated load in the centre with that produced by a uniformly distributed load of the same amount, and found the ratio of the one to the other to be practically the same by experiment as one would expect from theory. He made many other comparisons between theory and experiment on wooden beams with similar results.

Something of the same sort was done by Duleau with solid and built up beams of iron in 1820. Again the agreement between theory and experiment was satisfactory both for beams and for struts where the load at the onset of buckling coincided with that predicted by Euler from theoretical considerations. Duleau's tests on built-up beams—two flange pieces bolted together with spacers in between—showed deflections larger than he calculated, which was to be expected since he made no allowance for the effect of shearing force on the deflection. He concluded that it is advantageous to use beams consisting of two flanges joined by a web.

An important milestone in the progress of strength of materials was the publication in 1826 of a textbook on the subject by the French engineer Navier, who later became Professor of Mechanics at the École Polytechnique. This book sets out to show how the dimensions of structures can be calculated so that the deformations under given applied loads will not exceed the elastic limit, a marked advance over the practice of the previous century when theory and experiment were used to calculate the ultimate or breaking load. In his analysis of the bending of beams the author assumes that they have a plane of symmetry in which they are loaded and that cross sections remain plane during bending. Accordingly he concludes that the neutral axis passes through the centroid of the cross section and that the radius of curvature R of the beam is given by the product of the modulus of elasticity E of the material and the moment of inertia I of the cross section of the beam, divided by the bending moment M. Navier made numerous important contributions to the subject in his book, for example, in the study of statically indeterminate structures, the bending of curved bars and the strength of thin spherical shells. For this last case he made tests to confirm his theory using iron shells about one foot in diameter and 0·1 in. thick and found that the ultimate strength of the material was about the same for the spheres as for flat samples tested in a tensile testing machine.

In 1829 the French engineer S. D. Poisson made a discovery of fundamental importance to elastic theory, and as a result his name has been immortalized. He observed that for simple tension, within the elastic range, the axial strain must always be accompanied by a lateral strain, and that the two strains bear a constant relationship to one another, for a given material. The ratio of the lateral strain to the axial strain is still called Poisson's ratio.

Another important publication which appeared in 1833 was written jointly by two famous French engineers, Lamé and Clapeyron.

In it they showed that general equations could be used to determine stresses and deformations in many engineering problems of great practical interest. Starting from the consideration of the stresses at a point in an elastic body they showed by means of vectors how the magnitude and direction of the stress at any other point could be obtained. They applied their equations to obtain the correct solutions to a number of practical cases, for example, a hollow circular cylinder or spherical shell subject to uniform internal or external pressure, and the torsion of circular shafts and cylinders. Somewhat

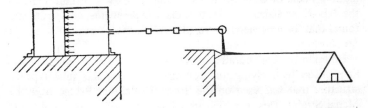

Fig. 128. Lamé's tensile test machine.

earlier, in 1824, Lamé designed and built a special testing machine (Fig. 128) which incorporated a hydraulic ram for producing the load and counter-weights for measuring it.

Engineering students will remember Clapeyron for his analysis of continuous beams, which led to the three-moment equation that he published in 1857.

One of the minor contributions of the English mathematician Sir George Cayley was his study of the strength of the bicycle wheel of today in 1808. He described it as a device 'to refer the whole firmness of the wheel to the strength of the rim only, by the intervention of tight strong cording'.

There seems to have been a shortage of English textbooks on strength of materials at this time, the most popular being by Thomas Tredgold whose book, *A practical essay on the Strength of Cast Iron* (1822), gave data for the design of cast-iron structures; another book by him entitled *The Elementary Principles of Carpentry* (1820) has a chapter on strength of materials; a third book by Peter Barlow called *An essay on the Strength and Stress of Timber* (1817) mentions experiments on the deflection of rails both under static load and when subject to rapidly moving loads as under a locomotive travelling at speed.

At the end of this period the leading English engineer on strength of materials was Eaton Hodgkinson, who in 1847 was appointed Professor of Engineering at University College, London. He made experiments with cast-iron beams and found that the most economical cross section was an I beam in which the lower (tension) flange had approximately six times the cross sectional area of the upper flange. The tensile strength of cast iron is approximately one-sixth of its compressive strength so that the practical result agreed again with theory. Hodgkinson was also among the first to observe and record that the fractures of his cast-iron compression pieces always occurred along a plane or conical surface having a nearly constant angle to the axis of the specimen. He, too, made experiments with struts and found that for long slender struts Euler's formula was confirmed but for short struts the formula did not apply.

Hodgkinson collaborated with William Fairbairn and Robert Stephenson in evolving and testing the rectangular tube type of structure that was used on the Britannia Railway Bridge over the Menai Straits. This was built up of wrought iron plates and joists into a section 28 ft. high and 14 ft. wide, the railway train travelling through the middle of the tube.

Another engineering professor to make some useful contribution at this time was Julius Weisbach at Freiberg in Germany. His work in 1848 concerned the design of machine parts that are subject to combined stress. For the selection of dimensions that would be adequate he used the maximum strain theory as a basis. Weisbach was also interested in the way engineering students should be taught strength of materials, and he is believed to have been the first to establish a laboratory where his students could experiment for themselves with beams, and make models of structures to verify their analytical work.

One of the phenomena that came into prominence as the result of operating the railways was that of the fatigue of metals. Among the first publications about it was one in 1843 by a young railway engineer W. J. Macquorn Rankine, afterwards Professor of Engineering at the University of Glasgow. Having examined locomotive axles which had unexpectedly failed after working satisfactorily for several years he asserted that the fractures began as minute fissures extending all round the neck of the journal and penetrating about $\frac{1}{2}$ in. in depth. He suggested that the fibres of the material had given way where there was a sudden change of cross section which prevented them from behaving elastically. He therefore proposed as a remedy that the journals should be made with a large curve at the shoulder. The fatigue of railway axles was also discussed at several meetings of the Institution of Mechanical Engineers in London in 1847, 1848 and 1849.

In 1848 a commission was appointed in England to inquire into the use of iron for railway structures. Large-scale fatigue experiments were undertaken at Portsmouth. A long horizontal bar was supported at each end and depressed by a cam situated near the centre of the span. Some specimens were subjected to as many as half a million depressions before fracture. The commission concluded that 'iron bars will scarcely bear the reiterated application of one-third of their breaking weight without injury'. To provide for safety William Fairbairn, a British designer, chose his dimensions on some of his bridges so that the ultimate load to produce failure would be eight times the heaviest load ever likely to be applied, or, as we should say now, he used a factor of safety of eight based upon the ultimate strength. Fairbairn was also interested in the strength of riveted joints for fastening iron plates together, not only for bridge construction but also for shipbuilding. In 1838 he found by testing that machine riveting gave considerably stronger joints than hand riveting.

TOOLS

The development of new machine tools and the further improvement of the old ones were essential factors in the industrial revolution that was taking place from 1750 to 1850. Not only were new general-purpose machine tools invented such as the milling machine, the shaper and the metal planer, but more significantly a new kind of machine tool made its appearance—the special-purpose machine tool designed specifically to perform certain operations on a particular part of a particular size, and suitable for no other purpose. A classic example of this was the machinery designed and built for the one purpose of making pulley blocks for the sailing ships of the British Navy, but there were others.

The machine tool which occupied the central position in these developments was the lathe. In 1750 it was still made of wood. The pole lathe with treadle was used for light work and for heavier work the lathe was driven continuously by a belt from a great wheel or a water wheel; for screw cutting, mandrels were used with threads at one end so that as the work rotated it moved longitudinally past the stationary tool as the thread was being cut. The alternative method of screw cutting by using a lead screw to move the tool axially had been proposed by Besson two centuries earlier but fell into disuse, probably because of the difficulty of mounting and working a lead screw satisfactorily in a wooden structure. Nevertheless, the lead screw was used by clockmakers, who also used change wheels and a metal frame for the so-called fusee engine, a special-purpose lathe used to cut the conical threads on the fusee for clocks (*see* Fig. 73). The introduction of the all-metal lathe, with lead screw, change wheels and compound slide rest is attributed to Maudslay, an English tool maker who about 1800 established in London the firm of Maudslay Sons and Field, which lasted for a century and served as a training ground for generations of famous British engineers and mechanics.[10]

Maudslay was the first to appreciate the essential importance of the accuracy of the guiding surfaces which control the motion of the tool during machining operations, and he used plane surfaces to do this. He had accurately scraped surface plates placed on the benches beside his workmen so that they could conveniently check the work they were doing. He was equally careful about the accuracy of the lead screws made in his works, and by continually checking and correcting the inaccuracies of his master screws, he eventually obtained after years of work, screws of much greater accuracy and length than any that had been produced before.

Some important improvements were made to the all-metal lathe in 1817 by Roberts,[11] an ingenious mechanic who devised the self-acting mechanism to make Crompton's spinning mule automatic and who was among the first to produce a successful planing machine. His additions to the lathe were the back-geared headstock, a four-stepped cone pulley for the drive, and a self-acting carriage driven by a long screw, which was not a lead screw for screw cutting, but simply for traversing the carriage for ordinary turning. It was driven by bevel gears from the main spindle through gears which could be used for disengaging and reversing.

By the end of the period the leading machine-tool manufacturer was Joseph Whitworth. His major contributions to the lathe were the hollow box-like form for the lathe bed (Fig. 129) with the lead screw in the middle; the split nut working on the lead screw that could be engaged by closing up the nut to make the saddle traverse; and the arrangement for working the cross feed from the lead screw. In this form the light general-purpose lathe became standardized for the next sixty or seventy years; indeed many lathes substantially of this design are still in use in all parts of the world.

For machining the inside of large cylinders a lathe was thought to be not so suitable as a boring machine in which all that had to be rotated was the cutting tool, mounted on the radial arm of a boring

Fig. 129. Whitworth's lathe.

bar in the centre of the cylinder. The early steam engines of Newcomen, Watt and Smeaton had very large cylinders—up to 6 ft. in diameter—and these had to be accurately bored to avoid leakage between the piston and cylinder. In 1775 Wilkinson, an iron master in Shropshire, devised for boring a cannon a horizontal boring machine that was able to do this (Fig. 130). Operated by a water wheel, its new feature was that the boring bar was hollow and carried a slot throughout its length, though it was stiff and heavy and was supported in bearings at each end. The disc carrying the tools was moved axially by a rod (which was not rotating) in the centre of the boring bar. The end of the rod, projecting from the machine, was attached to a rack which was moved axially by a pinion attached to weights and levers that could be adjusted to give the feed required to the tools. The cylinder being bored was lashed securely to a fixed timber cradle and the success of this machine was due to the rigidity of the fixing and of the boring bar which was supported at both ends.

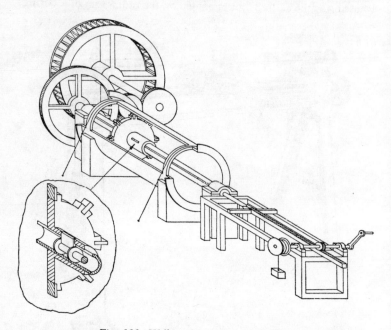

Fig. 130. Wilkinson's boring machine.

Before 1800 much of this boring work was being done on machines that were generally known as vertical boring mills. These machines were really lathes with vertical spindles and large rotating face plates to which cylinders were bolted for boring with fixed boring tools. Such a machine was installed in Boulton and Watts's Soho foundry in 1795 (Fig. 131). The advantages of this type of lathe—the vertical boring mill—for handling, setting up, and machining large cylinders and other components are too obvious to need explanation.

Copying machines of various kinds were developed during this period for copying medallions, panels of wood and stone, and even statuary. During his declining years, James Watt made some improvements to the machines then in use, but the copying lathe may be said to have come into general use in 1818 with the machine designed by Thomas Blanchard of Worcester, Massachusetts,[12] for turning the stocks of rifles and small arms. Some of his machines were later

Fig. 131. Vertical boring mill.

purchased for the Royal Small Arms factory at Enfield on the advice of James Nasmyth. The importance of Blanchard's achievement was that it was an essential step towards the more refined copying machines of today. In Blanchard's machine the pressure of the follower on the model was used to move the cutting tool, whereas in modern machines, a light detecting finger moving over the model can be used to actuate a servo-motor which moves the tool.

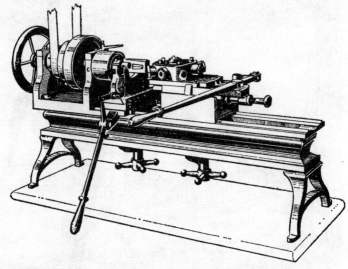

Fig. 132. Robbins and Laurence turret lathe, 1855.

One of the most radical changes to be made in the development of the lathe was the invention of the turret lathe. In the turret could be mounted a series of different tools for performing a sequence of operations on the work, and each tool could be brought into its correct working position by moving a single lever which turned the turret round from one station to the next. There is no doubt that turret lathes were in use in the United States before 1850 but they were not generally available on the market until 1855 (Fig. 132).

The planing machine was invented in England soon after 1800. Some say that Matthew Murray of Leeds was its originator and that he used it for machining the surfaces of D valves used in steam engines; others give the credit to Richard Roberts whose machine, built in 1817, is still in existence. Another early metal planer was built in 1825 by Joseph Clements, and was a large machine capable of machining work up to 6-ft. square, for which the charge made was 18s. per square foot. The machine was hand-operated, the table running on rollers, and there were tools for cutting during both directions of travel of the bed.

In 1836 a planing machine was built in the United States that had a

Fig. 133. Whitworth's planing machine.

bed of granite in which the cast-iron slideways were recessed and the table driven by a chain and sprocket wheel. Some of the early planers were mounted on the wall of the workshop. One of these, which was originally installed in James Watts's Soho foundry, can now be seen in the City of Birmingham Industrial Museum.

In 1842 Whitworth introduced a self-acting power-driven planing machine (Fig. 133). The table was supported in V guides and driven by a lead screw down the centre and rotated by bevel wheels to effect the reversing—a clumsy arrangement compared with the operation by open and crossed belts that superseded it. The tool box was swivelled round for cutting on the return stroke of the table, this motion and the crossfeed both being automatic. This machine, and a more modern type with a quick return motion, were both exhibited by Whitworth in the 1851 Exhibition in Hyde Park.

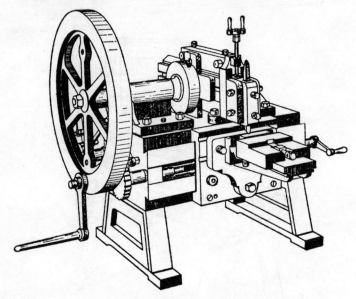

Fig. 134. Nasmyth's shaper.

The shaping machine was invented by James Nasmyth in 1836. At the time of Maudslay's death in 1831 Nasmyth was his personal assistant, but he then left the firm of Maudslay Sons and Field and set up his own workshop in Edinburgh and later in Manchester. Nasmyth's shaper (Fig. 134) is similar to many machines that are in use and even being manufactured today. The crank mechanism for giving a quick return to the tool was an improvement made soon afterwards by Whitworth.

Another machine that could be used for producing flat surfaces on metal was invented at this time—the milling machine. It originated in the United States in 1818 and was the work of Eli Whitney, the inventor of the cotton gin. The first milling machine made for sale was designed by Howe in the United States in 1848 (Fig. 135) but we have seen already that the process of milling had been used many

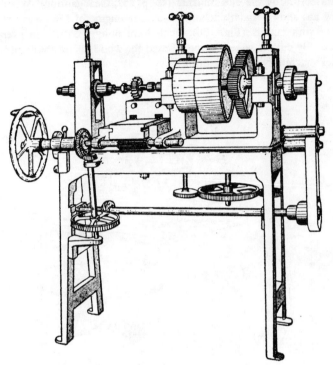

Fig. 135. Howe's milling machine.

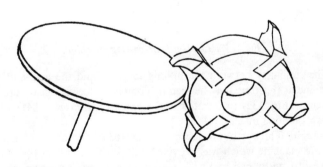

Fig. 136. Réhé's cutters.

years before by the clockmakers (see p. 152, Gear cutting). Woodbury has shown that Réhé's wheel-cutting engine of 1783 was provided with cutters (Fig. 136) which were milling cutters in every sense. Fly cutters were also being used for wood, as on the famous block machinery at Portsmouth.

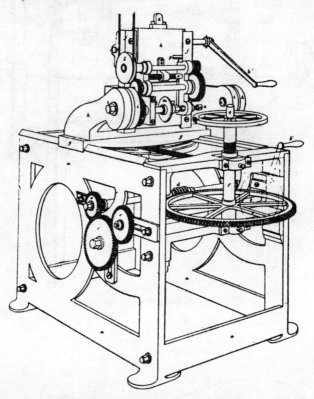

Fig. 137. Lewis's gear-cutting machine.

The similarity between the milling machine and the gear-cutting machine at this time may be seen by comparing Lewis's gear-cutting machine (Fig. 137) (illustrated in Buchanan's *Millwork*, 1841) with Howe's milling machine (Fig. 135).

Joseph Whitworth produced in the 1830s an improved gear-cutting machine which was provided with involute cutters that were power driven by belt and pulleys through a worm and worm wheel. This machine had geared indexing for the blank being cut.

The most surprising advance in gear cutting was the appearance before 1842 of a gear-generating machine due to Joseph Saxton of Philadelphia. The principle of the machine is shown in Fig. 138 from which it can be seen that the sides of the cycloidal teeth being cut are being generated by the cutter being rolled around the work

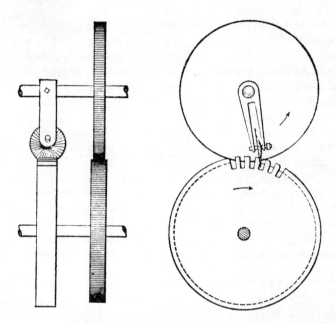

Fig. 138. Saxton's gear-generator.

at the correct radius to generate the correct epicycloidal shape. The serrations in the two wheels connecting the two shafts together were milled with very fine teeth so that they rolled together as the finished gears would do.

A machine tool that is to be found in every workshop is the drill. The vertical pillar drill with power drive and feed and adjustable table was in general use by 1850 and several of these machines and radial drilling machines were exhibited in 1851. In a somewhat cruder form the pillar drill had probably been in use since 1750, if not before. Flat arrow-headed drills were still in general use in 1850, for the twist drill did not appear for another ten years.

Most of the power-driven woodworking tools that we have today were in general use before 1850. The circular saw was used by Walter Taylor of Southampton before 1780; a rotary wood planer was built by Joseph Bramah for Woolwich Arsenal in 1802; in 1836 a machine was built for making tongues and grooves in floor boards using rotary cutters; and a successful bandsaw was made in the United States in 1849.

The spectacular results that could be achieved with power-driven woodworking machinery as compared with hand-work was well illustrated by the famous block-making machinery installed for the British Admiralty at Portsmouth in 1808. At that time the Admiralty had a requirement of 100,000 pulley blocks per annum for sailing ships. They were made by 110 skilled men by hand methods. Forty-three special-purpose power-driven machines were designed by Marc Brunel and built by Maudslay for a capital outlay of £54,000, as a result of which ten unskilled men were able to do the work with an annual saving of £17,000 p.a. The machines included circular saws, machines for milling slots with rotating cutters, boring machines, a mortising machine—which bears a marked resemblance to a vertical slotting machine—and a special lathe with a power-driven cross slide. The significance of this assembly of special-purpose machine tools is that they constituted one of the very first mass-production units.* Something of the same sort was done about the same time in the United States where the mass production of wooden clocks began in 1809, and by 1814 had caused their price to be reduced from $25 to $5. The mass production of clocks made in metal produced even greater savings so that the price many years later came down to half-a-dollar.

The mass production of interchangeable parts began much earlier in France where the parts for the delicate mechanism of the musket lock were being made interchangeable by the use of jigs in 1785. Thomas Jefferson, who was then United States Minister in France described how he had assembled a musket lock from components selected at random. The system of making interchangeable parts by mass production was developed independently in America where the method was rapidly improved from 1798 until 1835 when Colt revolvers were being made. By 1853 the armoury at Hartford, Connecticut, contained 1,400 machine tools to which the combined ingenuity of Eli Whitney, Simon North, Elisha Root and Colonel

* A number of these machines can still be seen at the Science Museum, London.

Colt had contributed. Jigs, fixtures, and gauges were a prominent feature of the methods used and their success in making interchangeable parts involved machining to accurate limits, accurate measurement, and above all accurate screw threads.

The pioneers in accurate workshop measurement and in the making of accurate screw threads were the British toolmakers Maudslay and Whitworth. Before 1830 Maudslay had made a bench micrometer accurate to 0·0001 in. that he used as a workshop standard (Fig. 139); by 1835 Whitworth had made a comparator for comparing yard standards of length to an accuracy of 0·000001 in.

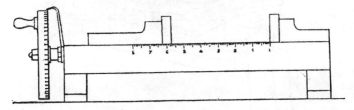

Fig. 139. Maudslay's micrometer.

i.e. one millionth of an inch. To use this he introduced the gravity or feeling piece which would slip with a movement of a millionth of an inch. In 1841 Whitworth was responsible for introducing standard screw threads which still bear his name and were standards in engineering for more than a century. He specified the numbers of threads per inch to be used for different diameters and the constant angle of 55 degrees between the sides of the V threads.

In his *History of the Grinding Machine*, R. S. Woodbury poses and answers some very interesting questions about the development of grinding, which has been throughout largely the work of the American toolmakers. In 1750 grinding was all done by hand on wheels of natural stone, or by lapping with emery. By 1850, the cylindrical grinding machine, the universal grinding machine, the surface grinder, the tool and cutter grinder and the disc grinder had all been invented and were in use in the United States and to some extent elsewhere. Further progress in the development of grinding was being delayed by the lack of suitable material for grinding wheels. It was realized that naturally occurring stones would not suffice to exploit the possibilities of the grinding process and efforts were being made to find suitable bonding materials to make synthetic wheels, but these were not successful until 1885.

In the eighteenth century grinding was used for finishing the external surface of rifle barrels. This was done by rotating the barrel while it was pressed against the circumference of a large sandstone wheel rotating at 400 rev./min. Machines of this kind—they were little more than grindstones—were used in the Springfield Armoury,

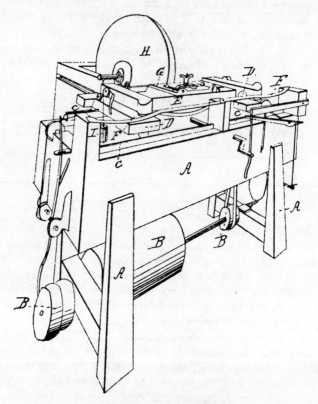

Fig. 140. Wheaton's elliptical grinding machine.

U.S.A., in Vienna and in Germany. One of the first real grinding machines was that developed by Wheaton of Providence, Rhode Island, in 1834 (*see* Fig. 140) which clearly stemmed from the lathe; but surprisingly it included an inclinable oscillating table on which the carriage was mounted for rotating the work, while the grinding wheel was separately driven about a horizontal axis which could be parallel to or inclined to the work. It was therefore a universal grinding machine—it could grind cylinders or tapers—and provision was made for fitting templates to which the carriage could be held by weights so that form grinding could be done.

Before this, Samuel Réhé in 1783 had invented a tool and cutter

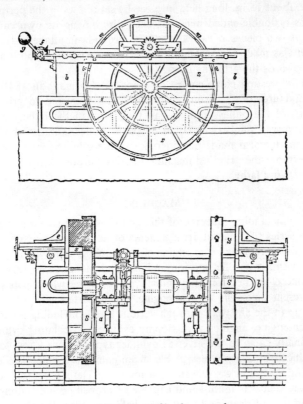

Fig. 141. Nasmyth's disc-grinder.

grinder which contained nearly all the features of its modern counterpart. The grinding disc was separately mounted in a headstock in front of the carriage holding the cutter, which was mounted between centres that could be swung round during grinding or fixed at any desired angle. Woodbury also mentions a patent issued in 1831 in Washington, D.C., to J. W. Stone for a surface-grinding machine which had a horizontal table on which the work was mounted and provided with a power traverse operated through a rack or screw. The grinding wheel had a vertical spindle and the pressure upon it was maintained by a spring. An interesting British contribution to grinding was made by Nasmyth in 1845 when he devised a disc-grinding machine (Fig. 141) having two cast-iron annular wheels, 7 ft. in diameter, divided into twelve segments each containing a stone about 15 in. long held in place by set screws in the periphery. It was a double-ended machine, each disc having its own carriage so that two pieces of work could be ground simultaneously. Provision was made for automatic feeding of the carriages both across and towards the grinding discs.

An interesting commentary on grinding was made in 1841 by a manufacturer of cast-iron belt pulleys who preferred to grind the surface of the pulleys rather than turn them in the lathe, because grinding was quicker: he could finish the pulleys thinner; however hard the metal it made no difference to the grinding machine, and furthermore the grinding machine for this purpose was cheaper to make than a lathe.

MACHINES

The outstanding machine of this period was the steam engine. A whole series of different arrangements of its parts was made, starting with the rocking beam engine of Newcomen, improved by James Watt with his parallel motion and sun and planet gear. Trevithick introduced the horizontal cylinder type of engine which persists to the present time. The vertical engine, which started with the crank-shaft up in the air, went through various forms, including the grass-hopper engine, engines with sloping cylinders, oscillating cylinders and finally came to the inverted cylinder type that we know today, introduced by Nasmyth with his steam-hammer engine. The most startling invention was the steam locomotive. Here was a self-acting machine that was mobile and capable of propelling itself along the ground and of drawing heavy loads behind it. The development of this machine included the consideration of many new factors—adhesion of the wheels to the track, suspension, the most suitable type of boilers, the arrangement of the cylinders—first vertical, then sloping, and finally horizontal—reversing gear, link motions for valve gear and, perhaps most important of all, some device for regulating the steam supply for varying conditions of speed and load. All these problems and many more were met and solved, sometimes most ingeniously, by the railway engineers of the day, for the most part in England, for locomotives and rails had quickly become important British exports to all parts of the world.

The mechanical requirements for successful locomotion on rails posed a number of fundamental engineering questions, the solution of which contributed to engineering generally. Those working on railway problems from 1825 to 1850 were in the very forefront of engineering progress, just as those working on nuclear engineering are at the present time. It was the locomotive engineers who introduced ring oil lubrication for bearings in 1827. They were also the men who tackled the problems of balancing rotary and reciprocating masses in locomotives in 1845, and the problems involved in accelerating, cornering and braking loads of hundreds of tons moving at speeds of a mile a minute were enough to tax the ingenuity of the cleverest engineers of any age.

One of the most remarkable achievements of British industry that contributed very largely to the industrial revolution from 1750 to 1850 was the mechanizing of the textile industry. This was brought about by a series of mechanical inventions of which the most important are now well known by the names of their inventors. Kay's flying shuttle, Arkwright's water frame, Hargreaves's spinning jenny, Crompton's mule, and Roberts's power loom, all belong to this period, and the inventive ability required to produce them was of the severely practical kind which owed nothing to science or mathematics.

Before the invention of the flying shuttle (in 1733) the shuttle in weaving was passed from one side of the loom to the other by hand or on the end of a stick. In Kay's invention the shuttle was thrown from side to side by the weaver jerking the cord attached to the picker as shown in Fig. 142. After passing between the warp threads the shuttle was caught and stopped gently at the far end without rebounding. The flying shuttle increased the speed of weaving by hand, but much more important than that, it was the essential invention that made the power loom a practical possibility. The development of the power loom proceeded in a number of steps during the next eighty years until in 1813 Horrocks of Stockport produced one of a type of which more than ten thousand were put into use during the next seven years. Soon afterwards this was superseded by the Roberts power loom (Fig. 143). It can be seen to consist of an assembly of two main horizontal shafts in a cast-iron frame together with gear wheels, cams, rollers, eccentrics, cranks, cords, springs, weights and levers, so arranged that all the operations of weaving were performed automatically. There was even an automatic stopping device to stop the machine if the shuttle got caught in the warp threads

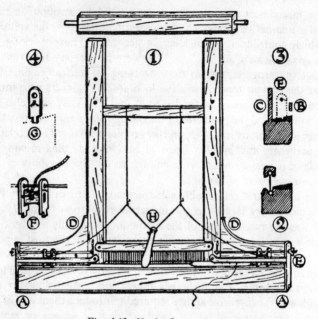

Fig. 142. Kay's flying shuttle.

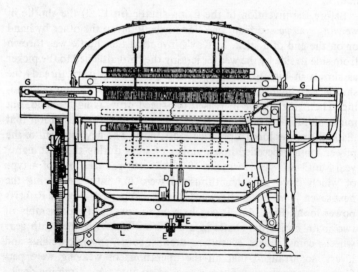

Fig. 143. Roberts's power loom.

composing the 'shed'. In all this no new machine elements were involved. By 1850 the power loom had superseded hand weaving for cotton and worsted cloths in England. Strassmann claimed that this had happened twenty years earlier in the United States.*

The Jacquard loom which came into use in 1804 was an important step in the development of machines because it was the first device

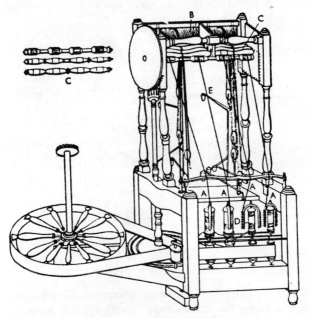

Fig. 144. Arkwright's water frame.

to be fed with 'instructions' by punched cards. The holes in the cards were used to 'instruct' the loom to weave a pattern which was determined by the number and position of the holes punched in the cards and the sequence in which they followed one another. The device of using punched cards to give instructions to machines was revived more than a century later for digital machining.

The mechanization of spinning was even more spectacular. By 1770 Arkwright's famous water frame was in use—so called because it was worked by a water wheel. The essential feature of the machine was that rovings of raw cotton were drawn out and twisted, by passing them successively between two pairs of rollers, the second going at a

* *See* Reference 2 in Chapter VII.

higher speed than the first, and then passing them down the arm of a flyer before being wound onto the bobbin. Arkwright's first machine (Fig. 144) had four bobbins. In so far as it was a continuous process Arkwright's machine was mechanically superior to those which followed it, but it was incapable of producing the finest types of yarn, mainly because the fibres were not twisted until after they had been drawn out. Hargreaves's spinning jenny was a closer approximation

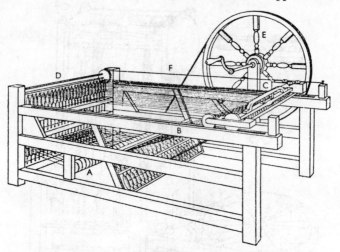

Fig. 145. Hargreaves's spinning jenny.

to doing by machine (Fig. 145) the same sequence of operations that had been done by the hand spinner. Many spindles were used to increase output and the rovings from all of them passed through notches in a clasp bar—replacing the thumb and forefinger of the hand spinner—mounted on a movable carriage so that the operations of twisting, drawing out and winding on could be performed intermittently. Spinning jennies of many different sizes were made in large numbers for more than fifty years, though before 1780 Samuel Crompton had made his first mule (Fig. 146)—so called because it was a cross between the jenny and Arkwright's throstle spinner—which was later to displace the jenny. In the mule the spindles were mounted on the movable carriage and the clasp bar was replaced by one or more pairs of rollers. Originally worked by hand, the mule was made self-acting by 1825 and has been in use ever since, particularly for producing very fine threads. Like the jenny it works intermittently.

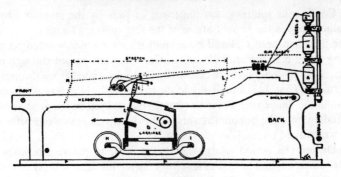

Fig. 146. Crompton's mule.

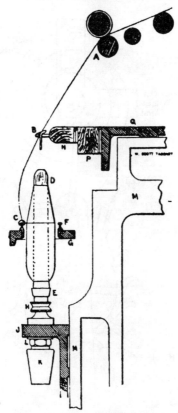

Fig. 147. Ring spinning-frame.

Continuous spinning was improved in 1828 by the invention by John Thorp in the United States of the ring spinning frame in which the flyer had been replaced by a small traveller which worked on a ring set in a frame around the bobbin. The yarn passed through a series of rolls—which were essentially feed rolls—and thence through a guide B (Fig. 147) and finally to the traveller C which was working round the flange ring F. As the frame rose and fell in the process of winding onto the bobbin the yarn was twisted and drawn out before it was wound on.

During this period efforts were being made to devise a machine for sewing cloth. The first patent for a sewing machine was taken out in 1790 in England by M. Saint but his machine was unsuccessful. So was that of Thimonnier patented in 1829. The modern type of high-

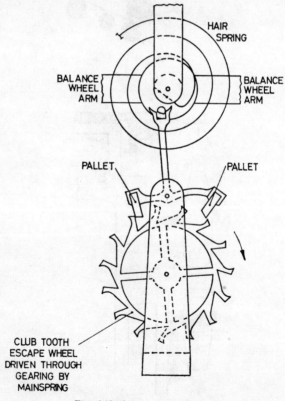

Fig. 148. Lever escapement.

speed sewing machine which quickly achieved success was patented
by Elias Howe in the United States in 1845.

By the end of this period the mechanical timekeeper had reached
its zenith in the detached lever escapement for watches and in
Denison's gravity escapement for tower clocks. Both owed their
success to the fact that they approximated closely to the ideal of a
'free' pendulum or escape wheel. In the lever escapement (Fig. 148)
the balance wheel was free of the train of wheels in the watch except

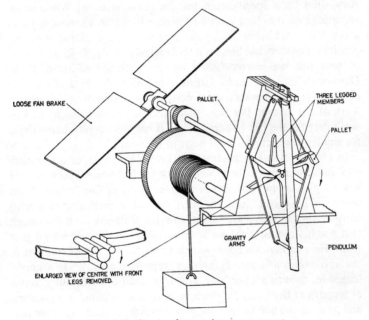

Fig. 149. Denison's gravity escapement.

during the short period of the cycle when it was locked and when it
was receiving its impulse. The Denison gravity escapement (Fig.
149) was even more sophisticated since the pendulum swung free
of the wheelwork the whole time and the two three-legged members,
with their three pins near the centre, were arranged to lift and lock
the two gravity arms instead of supplying an impulse to the pendulum
directly. Each gravity arm received its impulse from the central pins
in turn and then impulsed the pendulum as the arm fell under its
own weight and pushed the pendulum. The fly—a light two-bladed
fan brake loose upon its spindle—was provided to absorb some of

the energy released so that the three-legged members did not bang the pallets on the gravity arms too heavily when they brought the train of wheels to rest at each swing of the pendulum.

While the gravity escapement is one of the most beautiful pieces of mechanism that has ever been invented it is of course, mechanically inefficient, since much of the energy of the falling weight is dispersed by the fan brake.

It was about 1840 when the then Astronomer Royal, Sir George Airy, drew up a specification for the tower clock at Westminster requiring that the clock should keep time to within a second a day— a condition that many clockmakers of the day considered to be impossible. The story has been well told recently by Mr. T. R. Robinson, of how this was accomplished by the combined efforts of E. B. Denison (afterwards Lord Grimthorpe) and Mr. E. J. Dent, the clockmaker, who between them designed and constructed the clockwork of the famous Big Ben, which has run so successfully and so accurately for the last hundred years. It has been copied many times for similar tower clocks in all parts of the world.

In 1841 the first English textbook on the principles of mechanism was published by Robert Willis, F.R.S., Jacksonian professor of natural philosophy (physics) in the University of Cambridge. Willis seems to have combined the attributes of a mathematician with considerable mechanical insight and his textbook contains much that was his own original work—particularly in the sections on gear teeth. He said that his book was designed for the use of students in the universities and for engineering students generally. He acknowledged his debt to Monge, who fifty years before had begun a course of lectures at the École Polytechnique on the elements of machines, and pointed out that in 1834 the physicist Ampère had suggested that the name Kinematics should be given to that science concerned with motion (of machine parts) independently of the forces concerned. It was to be concerned with displacements, times, velocities and presumably the acceleration of the moving parts of machines. The four hundred pages of this book contain a very substantial proportion of the subject matter of a modern textbook on Kinematics, particularly on gear teeth—which occupies one-third of the book—gear trains, cams and linkwork. In the section on parallel motions he gives an analysis of Watt's parallel motion (Fig. 150) determining the proportions of the links from geometrical considerations and he gives a similar analysis for the link work of the grasshopper engine. He also discusses the sun and planet motion (Fig. 151) devised by James Watt

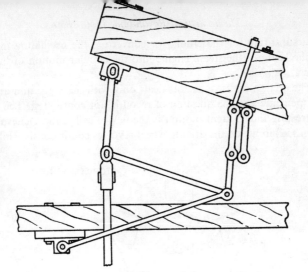

Fig. 150. Watt's parallel motion.

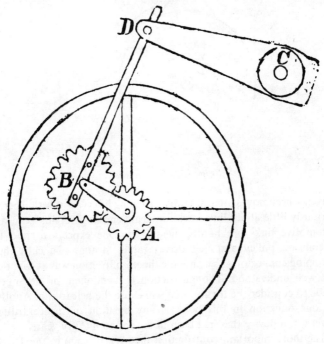

Fig. 151. Watt's sun and planet gear.

as a substitute for the crank, for converting the oscillating motion of the beam of the steam engine into the circular motion of the fly-wheel.

The section on escapements deals only with the verge and anchor escapements and the question of recoil is not considered. The most interesting intermittent motion is the device called the Geneva stop (Fig. 152) in which the driving wheel revolves continuously while the

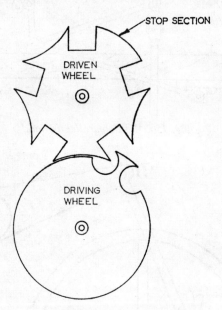

Fig. 152. Geneva stop.

driven wheel moves intermittently with long intervals at rest. There is curiously little about link-work for steam engine reversing gear, and locomotive links are hardly mentioned. One expects to find the section on pulley and belt drives rather strange—he calls them wrapping connectors—because the theory of friction was at that time not well understood. Though friction *is* mentioned, no attempt is made to consider the tension that will cause the belt to slip. Another curious omission in this book is any mention of the centrifugal governor, a device that had been in use for about fifty years.

The most important contribution that Willis made to the Theory of Machines was in connection with the shape of gear teeth, but to appreciate it we must examine what had been achieved before his time (as has been very well done by Woodbury). In the 1750's Leonhard Euler had published several mathematical papers setting out precisely the conditions that must be fulfilled for the satisfactory operation of gear teeth. He showed that these could be met by teeth having either an involute or a cycloidal profile, and he showed theoretically how these forms could be constructed. However, his work was too mathematical to be understood or even to be noticed by practical clockmakers and millwrights. According to Woodbury the first step in making the work of the geometers available to practical men was taken by Kaestner in 1781, because he showed a simple and convenient way of computing the shape of both cycloidal and involute teeth. As it was written in Latin it was not used by engineers of that day. The first books of use to them were Immison's *Elements of Science and Art* edited by Gill, 1803, and Hawkins's *Teeth of Wheels*, 1806. Hawkins gives reasons for preferring the involute form to the cycloidal, in particular that one involute wheel of a given pitch can work together with another of any size—except for pinions with a very small number of teeth—provided they are of the same pitch. This property is of great practical importance—for example for the change wheels of a lathe.

We come now to Willis's book of 1841.[13] This contained the first comprehensive analysis of gears of all kinds that was of direct use to the practical man. He introduced the constant $14\frac{1}{2}$ degree pressure angle for involute teeth—selected because this angle has a sine of nearly $\frac{1}{4}$. He pointed out that with involute teeth, backlash could be reduced by adjusting the centres of the wheels: he showed that the teeth of an involute rack have straight sides; but his greatest achievement was his invention of what he called the odontograph, a combined graphical and tabular method for laying out the profiles of involute teeth and of the cutters for making them. This method was much in advance of its day. Practical engineers continued to use epicycloidal teeth for years despite their disadvantages and it was not until sets of standard involute cutters were marketed by the Brown & Sharpe Co. in 1867 that the involute form for gear teeth became really popular.

Before the publication of Willis's book in 1841 Robertson Buchanan in his *Treatise on Millwork* had published data on the strength of gear teeth, both of wood and of iron with tables and charts for the use of designers.

The centrifugal governor for controlling the speed of revolution of engines and machines came into general use at this time. James Watt adopted it for his rotary engines and its invention is often incorrectly attributed to him. It is now known to have been used before his time for limiting the speed of rotation of the grindstones in windmills, but Watt adopted and improved it and was probably the first

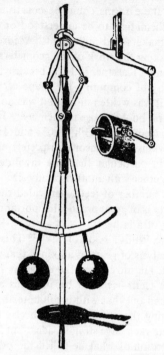

Fig. 153. Watt's centrifugal governor.

to use it as an automatic controller. In principle the same device (Fig. 153) is still used on prime movers that are required to operate at a constant speed—such as steam turbines in power stations—where it is important to maintain constancy of speed and hence uniformity of frequency of the electric supply because certain appliances such as synchronous electric clocks depend upon this for their correct functioning. Thus the accuracy of synchronous electric clocks depends upon the sensitivity of the centrifugal governor on the steam turbine at the supply station. As its name implies the centrifugal governor depends for its action on the centrifugal force of the rotating weights lifting them as the speed increases, and the vertical movement is used to operate the control valve of the machine.

Two other important machine elements began to appear in this period though it is doubtful if they were in general use, but both are illustrated in Willis's book. These were the differential gear* and the disc and roller drive. Yet a third remarkable invention of this period was the track type of land vehicle patented by Sir George Cayley in 1825. In this as in so many of his ideas he was far ahead of his time,

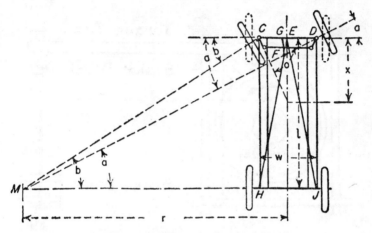

Fig. 154. Ackerman's steering-gear.

and it was not put to practical use until nearly a century afterwards. Another important machine element to appear was the metal-to-metal cone friction clutch which was used on several of the block-making machines designed by Sir Marc Brunel (1808).

The last four mechanical devices contributed later to making the mechanically propelled road vehicle a success, but a more important fact still was the adoption of the Ackerman steering gear (Fig. 154), an arrangement of the track arms on the stub axles and their connection by the track rod in such a way that the inner wheel moved through a greater angle than the outer one, so giving approximately true rolling of the wheels in cornering. The necessary condition for this is that in the plan view the point of intersection of the axes of all the wheels must meet at a point which always lies on the projection of the rear axle. In addition the front wheels are inclined, to bring the

* Cf. the Chinese south-seeking chariot (Fig. 37).

line of contact of the front wheels under the line of the pivots about which they turn when cornering. According to E. A. Forward* this mechanism was originally proposed by Du Quet for windmill carriages in 1714, re-invented by Lenkensperger in 1818 for horse-drawn vehicles and patented in England by Ackerman. Redmund used it in 1832 on a steam carriage.

In 1821 a new kind of machine was devised to meet a new need. This was a device that could absorb energy and measure the amount

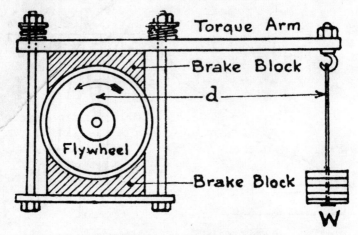

Fig. 155. Prony dynamometer.

absorbed. Known now as the Prony friction brake or dynamometer (Fig. 155) it is believed to have been invented by Piobert and Hardy in 1821 but it was first successfully used in 1833 for testing a small Fourneyron water turbine under the direction of Baron Riche de Prony who was then Director of the École des Ponts et Chaussées. Prony was awarded a prize for these tests and his name has been used to describe this type of brake ever since.

Some substantial achievements during this period took place towards making devices that could later be developed to make the control of machinery automatic. Such were James Watt's application of the centrifugal governor to regulate the quantity of steam supplied to an engine, the Jacquard loom and Blanchard's copying lathe which have already been mentioned, but this was not all. In 1794 Oliver Evans erected in the United States a continuous flour

* Science Museum Handbook—*Mechanical Road Vehicles*, part ii, p. 33.

mill in which the grain was not touched by hand from the time of its arrival to the despatch of the finished flour in bags. In 1833 the process of making ships' biscuits for the Royal Navy was mechanized. Lastly, as Professor R. H. Macmillan* has pointed out recently, a step of fundamental importance was taken in 1840 by the then Astronomer Royal, Sir George Airy, when he analysed the speed control of his telescopes and identified the phenomenon of 'hunting'. He showed that it could be expected to happen if the natural period of vibration of the regulating mechanism was related in a particular way to its speed of rotation. This work was far in advance of its time so far as engineers were concerned, and was not used by them, but it was a portent of what was to follow when automatic control became popular in the next century.

FLUID MACHINES

During this period we see the transition from the windmill and the water wheel as prime movers, to the steam engine and the water turbine. Most of the important discoveries concerning water turbines had been made before 1850. Yet the transition was not made suddenly. At the 1851 Exhibition water wheels were exhibited for sale and windmills were still being erected, but those who were in the van of progress had already tried out their ideas in practice and produced fluid machines that we still use today. Francis had built his inward-flow water turbine in 1840, Jonval had made his axial-flow water turbine in 1843 and by 1852 James Thomson, a brother of Lord Kelvin, had produced his vortex wheel with pivoted guide vanes, that could be coupled together and controlled by a governor.

The first important event of this period in the development of fluid machines was the publication by John Smeaton in 1759 of his famous paper to the Royal Society on the performance of water wheels and windmills. He built with his own hands a model water wheel 2 ft. in diameter and arranged it for test in an ingenious apparatus shown in Fig. 156. With this he carried out a lengthy series of experiments to measure the effect produced by the water velocity, the head of water and the wheel speed, with the undershot wheel shown in the diagram and with an overshot wheel of the same size, the apparatus being suitably modified to guide the water to the top of the overshot wheel. He found that the maximum overall efficiency he obtained with the undershot wheel was about 22 per cent and from

* Reference 17 in Chapter VII.

the overshot wheel about 63 per cent. The importance of these model experiments—and those he made on model windmills—derives from their being the first set of quantitative experiments made on scale models of full-size engineering machines.

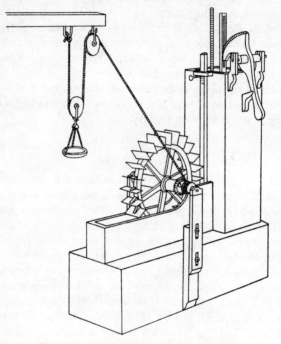

Fig. 156. Smeaton's water-wheel model.

Smeaton built a great many full-scale water wheels and used the results of his model experiments to guide him in design. One of the largest was an undershot wheel 32 ft. in diameter and 15 ft. wide built in 1768 for the London Bridge Waterworks on the Thames. Whenever circumstances allowed he preferred overshot wheels because of their higher efficiency. Smeaton's connection with the Carron Company's iron works, as consultant, led him to use cast iron in preference to wood in many of his designs, both of water wheels and windmills and he opened up a new era in power transmission machinery, thus bringing to an end a period of wooden construction that had lasted since pre-Roman times. The results of Smeaton's model experiments were widely used by designers of water wheels both in England and on the Continent. It was generally accepted that the best effect was obtained when the peripheral velocity of the wheel was about 200 ft./min. for medium-size wheels, but for wheels above 24-ft. diameter the velocity was raised to about 350 ft./min. Early in the nineteenth century a number of overshot wheels of 30-ft. diameter or more were built in iron by Joseph Glyn, an English engineer, who used these rules for his designs. One of the

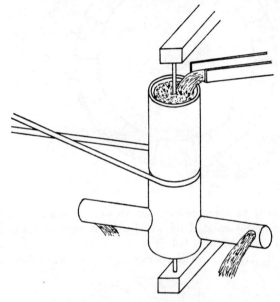

Fig. 157. Barker's mill.

largest water-wheel installations was that built in Scotland by William Fairbairn who coupled together four overshot wheels each 50-ft. diameter by 10-ft. width obtaining a total of 240 horse-power with an efficiency of 75 per cent. Another was a series of large water wheels installed at Niagara Falls in 1842 to supply power. However, the great majority of water wheels were very much smaller and according to Forbes* developed on the average between 5 and 10 horse-power.

About this time a reaction type of water wheel became popular for generating small amounts of power in country districts. Known as Barker's mill (Fig. 157) it operated on the reaction principle in the

* *See* the Bibliography to Chapter IV.

same way as Hero's aeolipile or a type of lawn sprinkler that is still in use.

Two other important innovations in the design of water wheels appeared at this time, the first being the breast wheel (Fig. 158), a

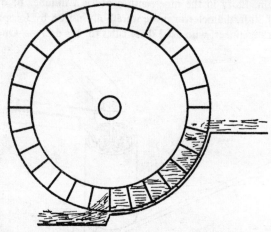

Fig. 158. Breast water wheel.

type in which the water comes in contact with the paddles at a height above the lowest point of the wheel, so that some of the potential energy of the water as it falls is used directly in doing mechanical work on the paddles. Smeaton constructed a number of these wheels

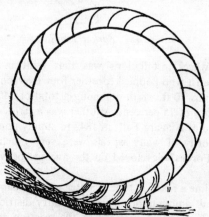

Fig. 159. Poncelet water wheel.

which had an efficiency somewhere between that of the overshot and undershot types. Early in the nineteenth century J. V. Poncelet made the important modification to the undershot wheel shown in Fig. 159. By curving the paddles as shown he provided for entry of the water into the compartments without shock, the idea being that the

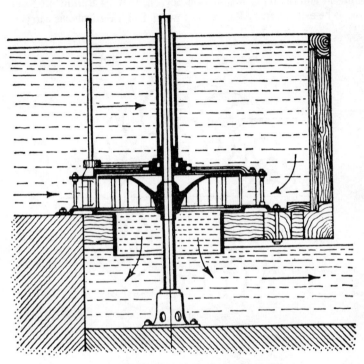

Fig. 160. Francis water-turbine.

water would run up the surface of the vanes and come to rest at the inner diameter and then fall away from the wheel with practically no velocity. This design raised the efficiency of the undershot wheel from about 22 per cent to about 65 per cent.

In 1826 Poncelet proposed an inward-flow radial turbine which was built some years later in New York. This was a vertical-spindle machine in which the runner with curved vanes was completely enclosed, the water entering the runner radially. Two years afterwards the machine was improved by the American engineer J. B. Francis who carefully designed the stationary guide vanes and the runner blades so as to provide for shockless entry to the runner and minimum exit velocity from it. Inward-flow turbines are now commonly referred to as Francis turbines (*see* Fig. 160).

The outward-flow water turbine was developed a few years earlier by the combined efforts of two French engineers, Professor Claude Burdin and his pupil Benoit Fourneyron. In 1824 Burdin proposed what he called a free efflux turbine which had a vertical axis carrying

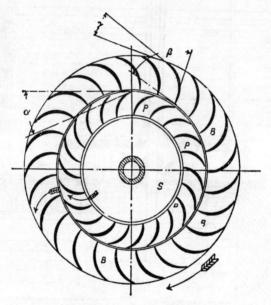

Fig. 161. Fourneyron's water-turbine.

a runner with curved blades through which the water issued almost tangentially. Fixed guide vanes, curved in the opposite direction, were mounted in an annulus inside the runner, an arrangement requiring considerable practical ingenuity to provide adequate support for the runner. Burdin was unable to surmount the practical difficulties involved in making a satisfactory working model but Fourneyron succeeded in doing so in 1827 (Fig. 161) and was awarded a substantial prize by the French Academy of Sciences. Subsequently he built more than a hundred machines of this type for all parts of the world. This type of turbine has achieved efficiencies of up to 75 per cent with a water head of anything from 1 to 350 ft., but such results can only be obtained at full power. The addition of a diffuser to the outlet—invented by Boyden in 1844—improved the efficiency about 6 per cent. This device was an annular casing surrounding the outer circumference of the moving vanes and of increasing cross section towards the outlet so that some of the kinetic energy of the water at exit was converted into pressure energy, thus increasing the effective

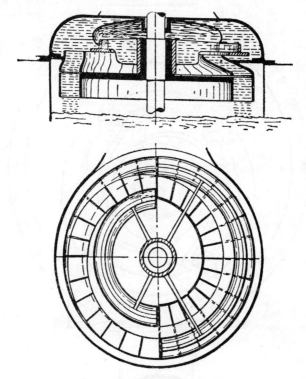

Fig. 162. Jonval water-turbine.

head. The outward-flow turbine is inherently unstable and difficult to govern, especially at part load, because the continually diverging passages result in loss of energy in turbulence and eddies. It was for this reason that the Jonval axial-flow turbine (1843) had a period of popularity. The water in passing through this turbine (Fig. 162) remained at about the same distance from the axis, so the flow was little affected by centrifugal force. The important improvement was to divide the wheel into concentric compartments, so that for governing at low loads the compartments could function separately. An even better scheme for controlling the part load performance was incorporated in Thomson's inward-flow turbine (Fig. 163). He employed pivoted guide vanes coupled together so that there was no sudden enlargement in the guide passages at any load, and accordingly the efficiency at part load was very little less than at full load.

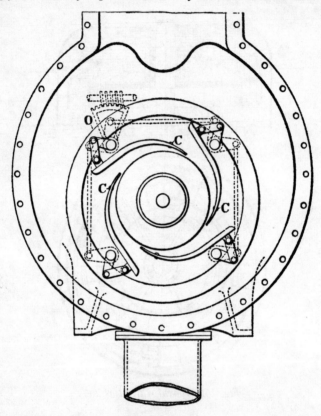

Fig. 163. Thomson's water-turbine.

Water began to be used in this period in a variety of devices for pumping and for supplying power to simple machines such as cranes, hoists, lifts, jacks and riveters. The simplest form of pump operated by water is the injector or jet pump in which a jet of water from a high pressure reduces the pressure in the area surrounding the jet and this can be used to raise water up a suction pipe and pass it to a delivery where it mixes with water from the jet.

The hydraulic ram pump was invented by the Frenchman J. M. Montgolfier in 1797 but came into general use about thirty years later. It had no cylinder or piston, only two valves in a valve box and an air vessel on the delivery line. Its action depended upon establishing a flow of water through the valve box and then interrupting the flow by suddenly closing a valve. Pressure waves were then set up and these continued to operate the valves indefinitely.

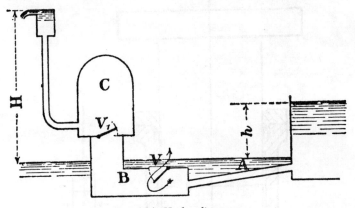

Fig. 164. Hydraulic ram.

Consider the diagram (Fig. 164). Flow is established in the first place by depressing the waste valve V. This is then suddenly closed, when the dynamic pressure wave set up will cause the delivery valve V_1 to open and the air in the vessel C will be compressed. When the energy has been expended the pressure falls and the water column momentarily moves back, opening the waste valve which will close again when the dynamic pressure is enough to overcome its weight. The cycle of operations then repeats itself. The hydraulic ram pump was used in hilly districts for supplying small quantities of water to a height above the stream in the valley. It was obviously not an efficient device.

Later some hydraulic air compressors were made, working on a somewhat similar principle but with mechanically operated valves.

One of the most important hydraulic machines to come into use in this period was the hydraulic press (Fig. 165). This was patented in 1796 by Joseph Bramah, a versatile engineer, who had patented a screw propeller in 1785, the modern water closet in 1778 and manufactured unpickable door locks and made some improvements to machine tools. To operate the press Bramah constructed a high pressure reciprocating water pump and the combination was used for a variety of industrial purposes, particularly for heavy forging when forces of several thousand tons were required to forge much larger pieces of steel than could be handled otherwise.

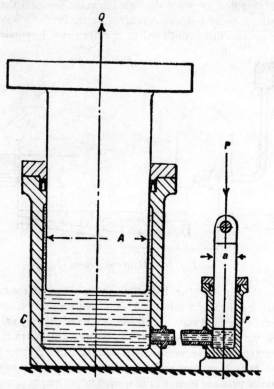

Fig. 165. Bramah press.

Another important application was the flanging of boiler plates while hot, to provide the dished ends of Cornish and Lancashire boilers. Bramah's press was the practical outcome of Pascal's 'machine for multiplying forces' (*see* Fig. 105) which in turn depended on the discovery by Simon Stevin that the pressure in a liquid is the same in all directions. Another application of this principle is the hydraulic jack (Fig. 166).

A significant contribution to understanding the resistance to flow of fluids in channels was made by the Frenchman Antoine Chezy who produced in 1768 the first formula for expressing the resistance to flow in terms of the velocity and the geometrical dimensions of the conduit. In 1797 the Italian physicist Venturi showed that by making a tube of two conical sections the water flowing through it would have the pressure reduced at the minimum section without the

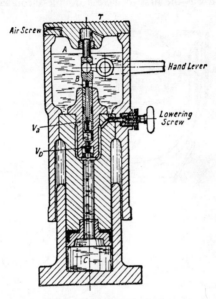

Fig. 166. Hydraulic jack.

formation of eddies, and the flowmeter later based upon this principle was given his name (*see* Fig. 222). Another contributor to hydraulic theory at this time was Coriolis who is chiefly remembered for his theorem concerning the analysis of the forces and components of acceleration in rotating systems, such as hydraulic turbines, in the course of which he identified what we now call the Coriolis component —the tangential component of the acceleration of the water due to its radial velocity.

The development of the centrifugal pump has been recently well reviewed by L. E. Harris[14] who showed that the centrifugal pump had now begun to challenge the reciprocating plunger pump for raising water, particularly for handling large quantities with comparatively low heads. Progress was made, principally in England and in the United States. Harris regards the year 1818 as a landmark in the history of the centrifugal pump because in that year the Massachusetts pump—Fig. 167—was introduced in America. Other designs were built by Blake, 1830, Andrews, 1839, and Gwynne, 1849, the last being 12 ft. in diameter, working at Pittsburgh. Blake's was a disc pump with semi-shrouded impeller while Andrews's improved pump of 1846 was the first to have a double-shrouded impeller. In

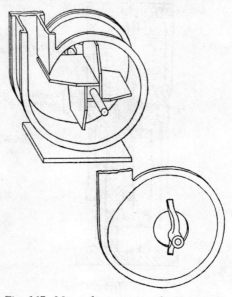

Fig. 167. Massachusetts centrifugal pump.

Britain Whitelaw, Appold and Bessemer were working independently to improve the centrifugal pump during the ten to fifteen years before the Great Exhibition in 1851, at which Appold, Bessemer and Gwynne exhibited centrifugal pumps that were submitted to independent tests. The results were:

	Quantity gal./min.	Head ft.	Speed rev./min.	Efficiency per cent
Appold	1,236	19·4	788	68
Bessemer	846	3·28	60	22·5
Gwynne	290	13·8	670	19·0

The principal factors in favour of Appold's pump were curved vanes on the impeller and the high speed. He later showed that with straight blades inclined at 45 degrees the efficiency dropped to 43 per cent and with radial blades to 24 per cent. The Gwynne pump was handicapped by its small size and by having only one blade, though soon afterwards more were added. The drawbacks of Bessemer's pump were its slow speed and its radial blades. It may be said that as a result of these tests, the design of centrifugal pumps became standardized for the next century or more.

Windmills continued in use, chiefly for milling flour, and many improvements were made to their design, but before 1850 it had begun to be evident that their hey-day was past. Experiments with a model windmill had been reported to the Royal Society by Smeaton in 1759 after he had reported his work on model water wheels. His model windmill had a span of 3 ft. 6 in., the sails being mounted at the end of a horizontal arm which was turned round so that the sails rotated in still air as the arm was whirled round a vertical axis. He found in what way the output varied with the wind velocity and his figures for the energy output from full scale Dutch mills agree quite well with modern measurements.

Only twenty-five years later, in 1784, the decline of the windmill may be said to have begun, when the Albion Flour Mills were built at Blackfriars, powered by two Boulton and Watt steam engines with cylinders 34-in. diameter by 8-ft. stroke. Though the mill was destroyed by fire seven years later, the undertaking was technically successful, with the result that many similar mills were built in the larger centres of population, causing the demand for small windmills inevitably to decline.

Improvements continued to be made to the design of windmills, particularly to the sails. In 1722 Meikle invented the spring sail, an arrangement of hinged shutters in the sail frame, controlled by a spring, so adjusted that when a pre-determined wind pressure was exceeded the shutters opened to 'lose' some of the wind. In 1789 Hooper devised a sail in which small roller blinds were fitted and arrangements were provided for adjusting all the roller blinds simultaneously while the mill was working. This was followed in 1807 by the so-called patent sail, a shuttered sail with hanging weights on the central chain so that the sail control was automatic.

The first balloon ascents were made in this period, and in 1785 the Frenchman Blanchard crossed the English Channel in a balloon with the American, J. Jeffries (Fig. 168). The possibilities of aerial flight by man had been discussed for centuries before this by such men as Leonardo da Vinci, Roger Bacon and Robert Hooke, but the first to achieve practical success were the Montgolfier brothers who were responsible for the first ascent in a balloon constructed of linen and paper containing hot air (with a capacity of 78,000 cu. ft.). The gas in the balloon was kept hot by burning straw and wool on a

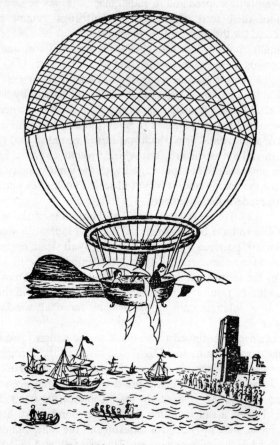

Fig. 168. Blanchard's balloon.

grate suspended some distance below it. Soon afterwards hydrogen gas, which had recently been discovered by Cavendish, was used and the heating discontinued. Thus began the ballooning era, which lasted for about 120 years. It was not a scientific pursuit but one of the activities of adventurous men, eccentrics, and showmen.

The scientific study of aviation with heavier-than-air machines began in this period with the outstanding intellectual achievements of Sir George Cayley. From the age of nineteen (in 1792) until shortly before his death at the age of eighty-four this gifted mathematician and scientist devoted much of his time to the analysis of and experiment with the problems of mechanical flight and the extent and importance of his contributions to aeronautics has never been surpassed. By the use of model gliders he found the weights that could be lifted by different surfaces and he showed why a cambered wing surface was more efficient than a flat one; he distinguished between the lift and the drag on an aerofoil, and realized that in flight there was a suction on its leading edge; he was responsible for the fixed wing, fixed vertical tail surfaces, for rear elevator surfaces, rudder surfaces, for streamlining to reduce drag and for the airscrew for propulsion. All this ensures Cayley's right to the titles of 'Father of Aerial Navigation' and 'Founder of the Science of Aeronautics'. He explained his interest in these words: 'An uninterrupted navigable ocean that comes to the threshold of every man's door, ought not to be neglected as a source of human gratification and advantage.'

HEAT ENGINES

During the whole of this period the reciprocating steam engine occupied the centre of the stage. In 1750 Newcomen atmospheric engines were being built in all essentials as they had been during Newcomen's lifetime—he died in 1729—whereas by 1845, a steam locomotive called the Great Western pulled a train of 100 tons from Paddington to Swindon at an average speed of 59 miles per hour.

The famous Englishman John Smeaton made a number of minor improvements to the Newcomen engine which resulted in the thermal efficiency being raised from its original 0·5 per cent to about 1·0 per cent. In 1774, an engine designed by Smeaton achieved a thermal efficiency of more than 1·1 per cent on test. It had a cylinder 52 in. in diameter and was built for pumping for Long Benton Colliery in Northumberland, being more carefully constructed than earlier engines, particularly as regards the cylinder, which had been bored in a special mill designed by Smeaton for the purpose, while the piston was packed with rope to minimize atmospheric leakage during the working stroke. Three years later Smeaton built an atmospheric engine of 60 horse-power at Kronstadt in Russia.

Owing to its design, the Newcomen engine was capable of doing mechanical work only during the down-stroke of the piston, that is,

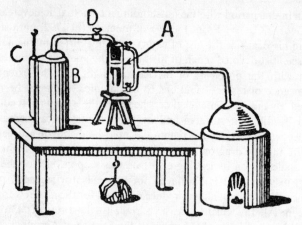

Fig. 169. Watt's experimental model.

it was a single-acting engine. Indeed its construction, with chain connection between the 'walking' beam and the piston rod, precluded the possibility of any upward thrust being exerted on the beam. To obtain rotary motion from the engine for driving machinery, it was used to pump water into an overhead tank from which a water wheel was driven, the water being continuously circulated. By this means very uniform rotary motion was obtained for driving cotton mills and other textile machinery.

A tremendous step forward in the development of the steam engine was taken by James Watt in 1769 when he took out his famous patent for the separate condenser. The romantic story has been told many times of how the young mathematical instrument maker at Glasgow University was given a model of the Newcomen engine to repair, and thought out, as he did so, why it consumed such an excessive amount of steam that it could not be kept running for more than a few minutes at a time. He realized, as no one else had done, that it was wasteful to subject the working cylinder alternately to a blast of hot steam and a spray of cold water, and that the cylinder should be kept as hot as possible the whole time, for much of the steam was being used up wastefully in reheating the cylinder every stroke. He therefore surrounded the cylinder with a steam jacket, abandoned the water seal and the exposed top of the piston, provided a cylinder cover with a hole and stuffing box for the piston rod, and used steam pressure instead of the atmosphere to force the piston down. The crucial in-

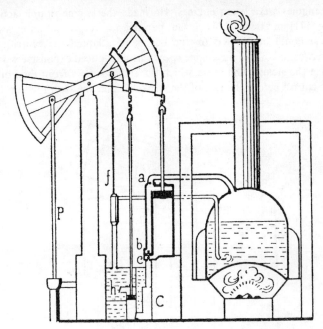

Fig. 170. Watt's single-acting steam-engine.

vention was to condense the steam, not in the cylinder below the piston, but in a separate vessel connected to it by a pipe with a valve between. When this idea first occurred to him he was not quite sure how quickly the steam would rush into the empty vessel to be condensed—or, as we should say now, how rapidly the pressure below the engine piston would fall—and so he made a model to try it out. The result was entirely satisfactory.

This model is of great historical interest since it was one of the first items of experimental apparatus made for a successful piece of engineering research. It is shown diagrammatically in Fig. 169 where A is the cylinder containing the piston, B is the condenser, C is the air pump for removing air before starting the experiment and D is the stop cock for admitting steam from the cylinder into the condenser. Immediately the stop cock was opened the piston rose and lifted the weight.

From 1775 to 1800, when he retired, James Watt made numerous improvements to the design of the condensing engine (Fig. 170). All his engines were beam engines. He made the engine double-acting (Fig. 171) and devised his elegant parallel motion so that an upward thrust could be exerted on the beam; he adapted the centrifugal governor to control the engine speed; he introduced expansive working in the steam cylinder in which the steam supply from the boiler was cut off before the end of the stroke so that the steam within the

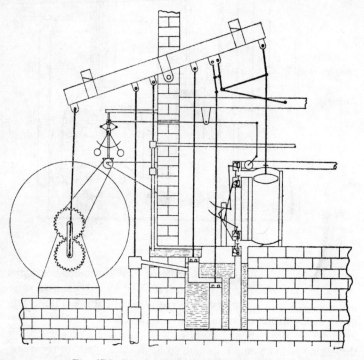

Fig. 171. Watt's double-acting steam-engine.

cylinder could expand and drive the piston before it; he used an air pump worked from the beam to withdraw air from the condenser; and when he was precluded by patents from using the crank to obtain rotary motion, he devised the sun and planet gear and used it successfully.

Watt combined to a most remarkable degree the attributes of the scientist and the practical engineer. It is not surprising that when these qualities were added to the financial resources and commercial acumen of Matthew Boulton that the firm of Boulton and Watt was extremely successful. In the period 1775 to 1800, 496 engines were installed by the firm; 308 were rotative, 164 were pumping engines, 24 were blowing engines.

The most enduring achievement of James Watt in the scientific sphere was to establish a unit for power, or the rate of doing mechani-

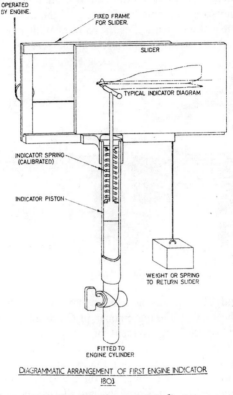

Fig. 172. Watt's engine indicator.

cal work. He coined the term Horse-Power (33,000 ft. lb. of work per minute) having established by experiment that an average horse could do two-thirds of this amount, i.e. he made his unit of mechanical power 50 per cent more. It is appropriate that we should now use his name for the electrical unit of power, the watt, which is the product of one volt and one ampere, though today we are more commonly concerned with thousands or millions of watts—kilowatts or megawatts (1 horse-power=0·746 kilowatts). Watt also determined the latent heat of steam and devised the first engine indicator (Fig. 172)—later improved by his assistant, Southern—and used it for setting the valves on his engine. He had a profound fear of high-pressure steam and would not use it, because of the danger of explosion and injury if the boiler or engine should burst under the pressure. Unfortunately such accidents were all too common in his day.

We cannot conclude this brief review of the development of the beam engine without mentioning the name of William Murdock, who spent the whole of his working life in the service of the firm of

Fig. 173. Trevithick's horizontal steam-engine.

Boulton and Watt, finally as Works Manager. He erected a great many of the firm's pumping engines in Cornwall and was directly responsible for the improvement of many of the engine details. Probably his principal contributions were the invention of the eccentric—a crank of small radius for the steam engine, the D-type of slide valve for the steam engine, and the use of gas for domestic and street lighting. Further, it is now thought that he suggested to James Watt the use of the sun and planet gear to obtain rotary motion from the beam engine.

After James Watt retired and his patents had expired the field was open for anyone to compete with the firm of Boulton and Watt in the design and manufacture of steam engines. The most prolific inventor to step into this gap was Richard Trevithick, a Cornishman who was the real inventor of the steam locomotive, the high-pressure horizontal steam engine, the Cornish beam engine and the Cornish boiler. Figure 173 is a side view of Trevithick's high-pressure engine with horizontal cylinder built in 1803. The cylinder was inserted into the top of the dome-shaped boiler but otherwise the engine, in most essentials, was little different from the horizontal type that persisted thereafter for nearly a century. The following year he wrote to a friend that nearly fifty of his engines were working, some with horizontal and some with vertical cylinders, being used not only for pumping water but for grinding corn, rolling iron and driving sugar mills.

The Cornish beam engines designed by Trevithick for pumping were first constructed about 1812 using a separate condenser and being similar to the Watt pumping engines in other respects except that they used steam at a much higher pressure (40 lb./sq. in. instead of 5). This enabled them to work with an earlier cut-off, about one-ninth of the stroke, so that a greater degree of expansion could be obtained. The steam valves were controlled by a 'cataract' device in which the passage of water through an orifice determined the instants at which the valves would be operated. For the purpose for which it was designed, as a pumping engine, the Cornish engine was unrivalled for nearly a century. The thermal efficiency of such an engine at Fowey with a cylinder of 80-in. diameter and a 10-ft. stroke, measured in 1834, was found to be 17 per cent (Fig. 174).

Another English engineer who helped to develop the high-pressure steam engine was Arthur Woolf, who added a high-pressure cylinder to an existing engine at the Meux brewery in London in 1803.* The practice of adding a high-pressure cylinder to a Watt beam engine so as to make it into a compound engine was later developed in 1848 by McNaught, who placed the second cylinder on the opposite side of the beam, providing a second parallel motion for it and using steam at a pressure of 150 lb./sq. in. as standard. As in all compound engines the steam from the high-pressure cylinder was exhausted into the low-pressure cylinder before passing into the condenser, and thus was expanded twice while doing work against a piston.

Trevithick's counterpart in America in the use of high-pressure steam was Oliver Evans, who constructed a steam dredging machine in 1803 and later built numerous vertical stationary steam engines working with high-pressure steam, and side-lever engines for the river boats of the Western rivers. By 1839 there were 250 of these engines driving the river paddle steamers, of which only eight were low-pressure, in spite of the risk attached to the use of high-pressure steam at a time when there was no satisfactory pressure gauge. They

---

* In 1781 a patent for the compound steam engine had been taken out by Hornblower.

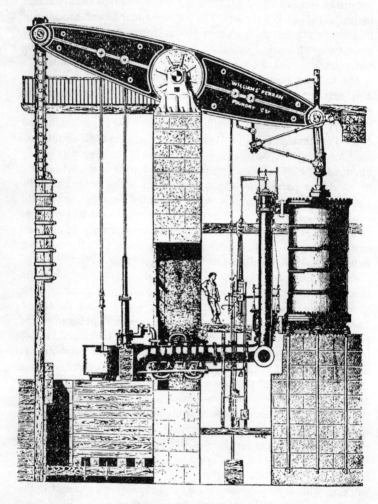

Fig. 174. Taylor's Cornish engine.

were preferred because they were more flexible and manoeuvrable, had lower capital and operating costs and, discharging their steam as they did to atmosphere, the difficulty of mud getting into the condensers did not arise. The boilers were, however, dangerous and many accidents occurred. By 1840, pressures of 100 lb./sq. in. were in common use, and in some cases 125–150 was often exceeded in an emergency.[15]

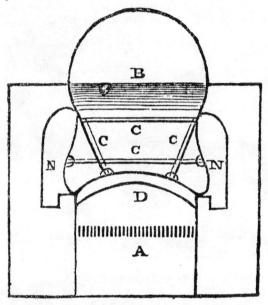

Fig. 175. Wagon boiler.

Next to the steam engine itself, its most important accessory—the steam boiler—was developed to a remarkable degree during this period. At the beginning it was little more than a brewer's copper directly fired from underneath, and having a lid of copper or lead to keep out the air. From this type of apparatus the haystack boiler and later the wagon boiler (Fig. 175) were evolved. Originally made of copper, they were by 1725 being made of wrought-iron plates and Dickinson quotes the cost of an iron boiler installed at Jesmond Colliery, Newcastle upon Tyne, as being £126 in 1733. This boiler, of 12-ft. diameter, was made of rolled iron plates riveted together and supplied (presumably) a Newcomen engine. During James Watt's time, it became necessary to increase the heating surface, for which the boiler was elongated to the form of the wagon boiler, a design inherently weak and unsuited to a pressure of more than about 5 lb./sq. in., though sufficient for the Watt engine.

Attempts to build boilers more suited to higher pressures were made simultaneously by Trevithick in England and Evans in the United States. Both men appreciated that a cylindrical cross-section was the most suitable shape, and so used one cylinder inside a larger

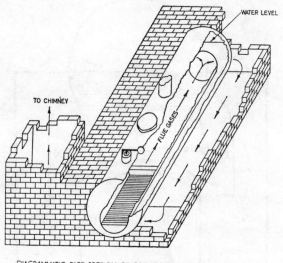

DIAGRAMMATIC PART-SECTION OF CORNISH BOILER

Fig. 176. Cornish boiler.

one. Trevithick placed the fire on a grate inside the smaller cylinder and thus invented the Cornish boiler (Fig. 176) which persisted as a satisfactory steam-raising unit for more than a century. Evans first used the smaller cylinder for the water, with the fire around it, but quickly adopted the same plan as in the Cornish boiler. The flat or rounded ends of these boilers were at first of cast iron, but later of wrought iron or steel. In 1845 Fairbairn patented the Lancashire boiler (*see* Fig. 238) in which there are two furnace tubes surrounded by a single shell. This was to remain in existence even longer than the Cornish boiler. The space occupied by the Lancashire and Cornish boilers is less than a third of that occupied by the old haystack boiler, for the same heating surface. Both types of boiler proved suitable for steam pressures of 200 lb./sq. in. and even more. The earliest Cornish boilers were 30 ft. long and 6 ft. in diameter, the central tube containing the fire being about $3\frac{1}{2}$ ft. in diameter, and extending the full length of the boiler.

The success of the *Rocket* steam locomotive (Fig. 177) in the Rainhill trials of 1829 (the first occasion of a competition between machines) was due very largely to the type of boiler designed by the

Fig. 177. The Rocket.

Stephensons, who used a horizontal fire-tube boiler containing a number of horizontal tubes extending from the fire box through the water and discharging the smoke into a smoke box below the funnel. Another essential feature that helped towards the success of the *Rocket* was the exhausting of the steam from the engine cylinders through a pipe in the smokestack so as to create an increasing draught as the speed of the engine increased. The idea of doing this was due to Richard Trevithick, who used it in the small locomotive exhibited in London in 1808 (Fig. 178) and in his Pendarren locomotive in 1804. Trevithick had also been the first to demonstrate that the adhesion between the locomotive and the rails was sufficient for the engine to pull a heavy load of trucks from a standstill without using a rack. There seems little doubt that George Stephenson adopted these ideas from Trevithick, probably picking them up from one of his workmen during the time a locomotive was being constructed to Trevithick's design at Gateshead in 1805.

Fig. 178. Trevithick's portable locomotive.

George Stephenson built his first locomotive in 1813 for use at Killingworth Colliery, Northumberland. It would seem that his achievements as 'father' of the railways were concerned more in solving the many problems involved in laying out the railway, such as choice of rails, supports, signals, tracks, bridges, viaducts and so on, rather than by the exercise of any exceptional mechanical originality. He was an excellent judge of existing mechanical devices and unhesitatingly chose the right one for the purpose, adopted the ideas of others to produce a successful complete machine in the various locomotives that were designed by him and his son, Robert. The romantic story of the development of the early railways from the Stockton and Darlington Railway onwards (1825) has been well told by Samuel Smiles and more recently by L. T. C. Rolt.

The transformation of the locomotive from the crude unsprung form of the lumbering *Locomotion*—the first locomotive that ran on the Stockton and Darlington Railway in 1825—to the almost modern design, as seen in the Harvey Combe locomotive on the

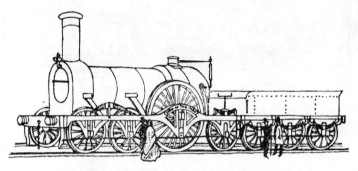

Fig. 179. The Lord of the Isles.

London and Birmingham Railway in 1835, was, according to Mr. L. T. C. Rolt, entirely due to the work of Robert Stephenson. He made improvements in each successive design, viz.—the *Lancashire Witch*, 1828, the *Rocket*, 1829, the *Northumbrian* and the *Planet*, 1830, the *Patentee*, 1832, and finally, the *Harvey Combe*, 1835, the first three having sloping cylinders and being directly driven by connecting rods fastened to the driving wheels, whereas the last two had horizontal cylinders with cranked axles. All these locomotives had to be fired with coke since they were required 'to consume their own smoke'. The great reputation as a civil engineer that Robert Stephenson had acquired by the end of his life, coupled with the aggressive personality of his father, who over-shadowed the earlier part of his career, continues to obscure the important contribution to the history of mechanical engineering that Robert made in developing the steam locomotive.

One of the most famous locomotives before the end of the period was the *Lord of the Isles* (Fig. 179) a triumph of locomotive design for the Great Western Railway, whose chief engineer—I. K. Brunel—though a man of personality and power, seems to have contributed

Fig. 180. The Great Eastern.

little of importance to the history of mechanical engineering, apart from his two classical blunders, his advocacy of the atmospheric railway, and of the broad gauge line of seven feet. The atmospheric railway has been described as a rope railway having a rope of air. A 15-in. pipe was laid between the rails and the front carriage connected to a piston in the pipe by means of an arm that passed down through a slot that ran along the whole length of the pipe, which was sealed by leather strips along the slot. The front of the piston was exhausted by a series of stationary pumping engines, positioned at intervals of a few miles along the railway line. The principal weakness of the system lay in the leather seals, which froze in winter and were eaten by vermin when greased, though there were other difficulties of corrosion, lubrication and weathering. Had there been complex trackwork or marshalling yards, which did not exist at that time, the atmospheric system would have been seen earlier to have been impossible to work. In the event, it was one of the most costly failures in the history of engineering at that time, for example, a renewal of the leather strip from Exeter to Newton Abbot cost £25,000, and when the system was finally given up the sale of machinery at scrap prices realized nearly £43,000.

In a different sphere we cannot but admire Brunel's courage and foresight as he designed and equipped his three great ships, the last

Fig. 181. Paddle-steamer on Mississippi.

of which—the *Great Eastern* (Fig. 180)—showed features in the design of the equipment that were far in advance of its time. The first of these ships, the *Great Western*, was the first true Atlantic steamship, and the second, the *Great Britain*, was the first ship to be driven by a screw propeller and the first iron ship.

Other particulars were as under:

	Great Western (1837)	Great Britain (1845)	Great Eastern (1859)
Tonnage	1,340	3,400	21,000
Steam pressure lb./sq. in.	5	15	25
Indicated horse-power	750	1,500	8,300

Some of the most powerful steam engines used in the United States were those of the paddle steamers plying on the Western rivers (Fig. 181). L. C. Hunter[15] observes that in the late 1840s the average power of the engines of boats on the Ohio River was about

1,200 h.p. They were slow speed, high-pressure horizontal direct-acting engines with cylinders of 30-in. diameter and an 8-ft. stroke, and thus were much larger engines than the stationary ones in use at that time in industrial centres in the U.S.A. For example, in 1838 only 10 out of over 200 engines in Philadelphia were rated at more than 50 h.p.: likewise, at the same date nearly two-thirds of the 200 engines in Pittsburgh factories and mills were less than 20 h.p., the largest units being for rolling mills, where the engines were 140 h.p.; even the railway locomotives in America had very small engines, seldom more than 25 h.p. By all these standards the power of the steam-boat engines—1,000 to 1,200 h.p.—was very large. Great risks to life and property were taken to achieve such concentration of power. The boilers had 4, 5 or even 6 flues instead of the two of the Lancashire boiler, and the sides were usually flat instead of circular, with the result that they were seen to 'pant' visibly with the opening and closing of the engine valves, and it was not uncommon for the engineer to insert wooden wedges in the joints before raising steam.

Two of the most dangerous features of these early high-pressure boilers were the absence of any satisfactory means of measuring the water level and of measuring the steam pressure in the boiler. Moreover, water was supplied to the boilers only when the engine was running (a defect that was shared by the early locomotives). The water level was measured by try-cocks positioned at different heights on the boiler shell, but as late as 1850 there were no reliable steam gauges. (The Bourdon bent-tube gauge was not invented till 1845.) The boilers were not provided with artificial draught so, to obtain a sufficient natural draught, the chimneys were raised higher and higher as power requirements increased, until by 1850 the tops of the chimneys on the larger steam boats were sometimes 90 ft. above the surface of the water.

During most of this period great efforts were being made to design a practical steam turbine but this was not achieved until Parsons produced his reaction machine in 1884, the fascination of the problem being such that between 1784 and 1884, nearly two hundred British patents were taken out for steam and gas turbines.

One of these early patents was granted to Baron von Kemperlin in 1784, and was brought to the attention of James Watt by his partner Boulton, who expressed fears as to the effect of the proposed turbine—a reaction wheel—on their business in reciprocating steam engines. After making some calculations Watt replied:

'The success of the machine depends upon the possibility of producing prodigious velocities, in short, without God make it possible for things to move at a thousand feet per second, it cannot do us much harm.'

In the same year, 1784, James Watt himself took out a patent for a reaction turbine in which the high velocity was avoided by using the steam pressure to force a liquid—mercury, water and oil are mentioned—out of tangential openings in the base of a rotating vertical cylinder which was divided along its axis into two equal chambers to be alternately filled and emptied of the fluid as the cylinder rotated. It is curious that James Watt, the inventor of the separate condenser, should ever have suggested bringing steam and water into contact in the machine, knowing as he did the inevitable waste of steam that this would involve. One must conclude that he intended it to work with mercury or oil and that he had been inspired by a popular hydraulic machine of the time known as Barker's mill (see Fig. 157).

Among the many patentees of improvements to the steam turbine before Parsons were Trevithick of steam carriage fame in 1815; Ericsson, 1830—whose name has been immortalized by the hot air engine cycle that bears his name, and the author's ancestor, Timothy Burstall, better known for his steam locomotive the *Perseverance*, which competed unsuccessfully against Stephenson's *Rocket* in the Rainhill trials of 1829. Burstall's patent for steam turbines dated 1838 was concerned with the Hero type of reaction machine and was notable for the suggestion that a set of arms curving in the opposite direction to the main ones should be mounted on the same shaft for the purpose of reversing the direction of rotation. There are many fascinating features mentioned in the patents that followed. Heath, in 1838, described a Hero turbine with trumpet-shaped nozzles; Gilman, the year before, had described a multicellular radial-flow turbine in which the steam was to be expanded in stages; Pilbrow, in 1843, experimented with 'the impulsive force of steam issuing from nozzles', and calculated that the best velocity for the vanes of the turbine wheels was about 1,250 ft./sec. For locomotion on land he proposed to use an air propeller directly coupled to the rotor of a steam turbine, in the stationary casing of which there were fixed vanes or guides 'to reflect the steam back upon the wheel for a second or other number of impulses'. Pilbrow's multiple-effort turbine was intended to operate by what is now called 'velocity compounding'. The man who came nearest to anticipating the invention of 'pressure compounding' for which Parsons was granted a patent in 1884, was Robert Wilson of Greenock who, in 1848, obtained a patent for steam turbines both of the radial-flow and axial-flow types and having both fixed and moving blades. Pressure compounding is a term used to describe dividing up the total drop of pressure of the steam among a number of nozzles and wheels in a series of small stages. None of these inventors achieved success.

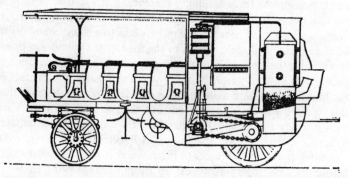

Fig. 182. Hancock's steam-carriage.

Some measure of success was achieved in a different direction however, when T. Hancock, using the reciprocating steam engine, introduced his famous steam carriages (Fig. 182) for road traction.

Heat engines using air instead of steam as the working fluid came into use soon after the beginning of the nineteenth century. Some of the terrible boiler explosions that occurred about this time probably led a number of thoughtful people to seek a means of substituting air for steam by using the gases from a furnace directly in the engine cylinder, thus disposing of the boiler altogether. The first to do this was Sir George Cayley in 1807. Little is known of the details of his first engine (Fig. 183) but he continued to improve it and in 1837 took out a patent for a better design which he claimed would produce five brake horse-power when consuming twenty pounds of coke per hour.

Meanwhile two other inventors had devised independently two other solutions of the problem of producing power from heated air. The first was the Reverend Dr. Robert Stirling who in 1816, when he was only twenty-six, patented the closed-cycle external combustion engine with regenerator in which the construction and use of a regenerator was first described, as well as the proposal to use a closed cycle.

274

In the closed cycle the same quantity of air—or working fluid—is used in the engine over and over again indefinitely, being passed successively through the processes of the cycle—heating, expansion, cooling, compression—without being renewed. In the open cycle a fresh quantity of air is admitted at the beginning of each cycle and discharged at the end of it.

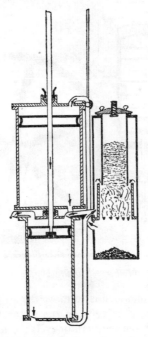

Fig. 183. Cayley's hot-air engine, 1807.

The regenerator was a quantity of material that could absorb heat from the air at one part of the cycle and give the heat back to the air when required at a later time during the same cycle.

Stirling was the first to propose the use of a regenerator in an engine, and he also foresaw its usefulness in the smelting of iron and in glass making. Stirling's name has been associated particularly with the closed-cycle type of heat engine incorporating a regenerator in which heat is added to the working fluid while it expands at a constant high temperature and rejected while it is compressed at a constant lower temperature. The working fluid is maintained at constant volume while heat is being stored in the regenerator and again while the regenerator is giving up heat.

Figure 184 shows a diagrammatic view of the engine reproduced from the patent specification. It is a single-cylinder inverted-beam engine with two rocking beams operating two pistons, 2 and 9, within the cylinder. Two is the cold piston driving the engine, 9 is a dis-

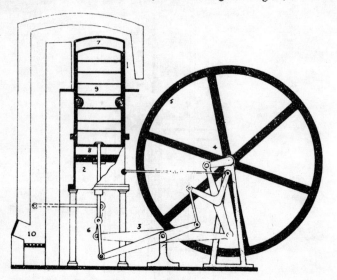

Fig. 184. First Stirling engine, 1816.

placer, the function of which is to move the air from the hot space 7 down the annular space between the piston 9 and the cylinder walls and so through the regenerator coils located in the middle of the displacer. The top of the cylinder is heated by furnace gases issuing from the fuel burning on the grate 10. The lower part of the cylinder is kept cold. An engine of this type was used for pumping water in a stone quarry in Scotland in 1818 and is believed to have produced 2 h.p.

Robert Stirling and his brother James produced a number of improvements in many engines they designed between 1820 and 1850. One of the largest, of 45 h.p., was installed in a foundry in Dundee in 1843. In this engine the cylinder was inverted so that the cylinder cover was directly above the fire where it must have been maintained at near a red heat, for the air was heated to 600° F. It is not surprising to learn that three of these cylinder covers were burnt out in four years' working, and that the lack of suitable metals to withstand such high temperatures prevented the air engine succeeding commercially except in quite small sizes, though the fuel consumption was less than half that of the steam engines of the time.

The other notable inventor of hot air engines was the Swedish engineer John Ericsson, a prolific inventor. He is best known for the closed-cycle heat engine with constant pressure, heating and cooling, applied by Joule in its reversed form for refrigeration and therefore called the Joule cycle. Most of the hot-air engines built by Ericsson worked on the open cycle principle described in one of his early patents of 1826, taken out soon after his arrival in England. In 1839 he moved to America where he continued building experimental engines which were finally successful after some spectacular failures. The worst of these were the four hot-air engines built to propel the paddle ship *Ericsson* in 1853. Ericsson calculated that the engines would provide 600 h.p. and consume eight tons of coal per day. The engine cylinders were 14 ft. in diameter and the stroke of the pistons 6 ft., speed 9 rev./min. During the trials the power was found to be half the design figure and the fuel consumption double. Modifications and redesign made no improvement and the engines were replaced by steam engines. In spite of this failure Ericsson continued his development on air engines and eventually produced a small engine of about $\frac{1}{2}$ h.p. which was so successful that by 1860, 3,000 had been sold and were in regular use in many countries of the world in situations where a small reliable engine was required.[*] The principal drawback of this type of engine was its great bulk and weight, even for quite small outputs. It also took some time to warm up—usually about two hours from lighting up until full power was obtained. One of these engines has been preserved in the Science Museum, London.

One of the most surprising achievements of this period was the invention of the vapour compression mechanical refrigerator by Jacob Perkins in 1834. The diagram of the apparatus, Fig. 185, shows that the compressor P was provided with automatic valves and delivered the compressed vapour into a coil of piping surrounded by cold water, where it condensed into liquid before passing through the throttle valve D on its way to the evaporator, which is the space between the two bottoms of the freezing basin A. The vapour was then returned to the compressor. Here we have all the essential features of the modern compression type of refrigerator. Jacob

[*] *See* the reference for Finkelstein in the Bibliography to Chapter VII.

Perkins was ahead of his time with this invention and it was not until 1857 that the process came into general commercial use. The vapour used was a hydrocarbon obtained from the destructive distillation of rubber.

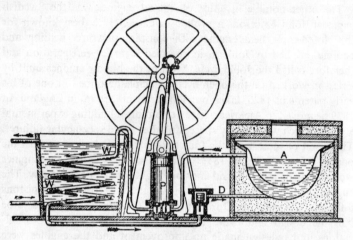

Fig. 185. Jacob Perkins's refrigerating machine.

Perkins certainly appreciated the importance of using as a working fluid one that could be condensed, evaporated and continually recirculated at suitable temperatures and pressures; whether he also realized that he had produced the first heat pump or reversed heat engine remains obscure, since it was not until after 1850 that Lord Kelvin and Professor Rankine became interested in the thermodynamics of refrigeration.

REVIEW

This was the period of the industrial revolution and for this the steam engine was partly responsible. It can surely be no accident that the country where the steam engine had been developed—Britain—was also the place where many machine tools and textile machines were invented, where the making of cast iron and wrought iron was cheapened, where mechanical clockwork was perfected, mechanical refrigeration was applied and, perhaps most important of all, the country where the railway first became a reality. We have followed the sequence of these events during the years between 1750 and 1850. James Watt transformed the steam engine from the crude water pump made by Newcomen into an efficient prime mover providing rotary motion to supply power for working factories, textile mills, and other purposes on land. Others followed him but to a great extent they built upon his work, so that James Watt should be justly awarded the title of the 'Father of Mechanical Engineering'.

One measure of the performance of the steam engine is its thermal efficiency, for which the following improvement was recorded:

Date	Type	Thermal efficiency (*per cent*)
1750	Newcomen	0·5
1767	Modified by Smeaton	0·8
1774	Further modified by Smeaton	1·4
1775	Watt	2·7
1792	Watt Expansive	4·5
1816	Woolf Compound	7·5
1834	Trevithick Cornish	17·0

James Watt did much more than improve a heat engine; he devised new mechanisms—such as his parallel motion—and made them work, and perhaps most important of all he applied a governor to his engine so that it could work automatically, notwithstanding variations in the load to be supplied; by so doing he also became the first control engineer.

Before Watt retired in 1800 all the engines built by Boulton and Watt—who had a monopoly—were beam engines, but very soon afterwards Trevithick introduced his horizontal high-pressure engine (Fig. 173), Maudslay brought out a vertical table engine with the crankshaft in the air and, after a period of groping for the most suitable form, Nasmyth designed his vertical steam hammer engine which became a standard form for the rest of the century.

It is interesting to notice how quickly the steam engine was applied to transport. The firm of Boulton and Watt had supplied some condensing engines for river boats, but for transport by land the condenser was a handicap. Later on, when Trevithick had developed the high-pressure steam engine, he applied it to give demonstrations on a miniature railway (Fig. 178), to show its possibilities. That the railway gained general acceptance after 1829, after the early opposition from landowners and vested interests, was in a measure due to the publicity given to the famous Rainhill trials (aptly described by L. T. C. Rolt as the first competition between machines). By 1850 when the railway mania of speculation had begun to subside, railway transport for goods and passengers was firmly established and was spreading rapidly throughout the world. The rapidity and success with which the steam locomotive and the railway were developed in the next fifty years is astonishing. Credit for the early pioneering work must go to George and Robert Stephenson, to I. K. Brunel and his locomotive engineer Daniel Gooch, designer of the *Lord of the Isles* (Fig. 179).

The successes that were achieved were not unmixed with failures. Lack of knowledge, foolhardiness, and in some cases sheer stupidity or carelessness led to terrible accidents, as from boiler explosions. Brunel made bad errors of judgment over the atmospheric railway, and even James Watt put forward in a patent for a steam turbine, proposals that were in conflict with the principle of his own separate condenser. Many failed in attempts to invent the steam turbine. The gifted Swedish engineer, Ericsson, who eventually produced a small hot-air engine that was very successful, made some appalling miscalculations of the power of the engines for the ship that was named after him.

Some very important advances were made in fluid machines. Smeaton's experiments with scale models of water wheels and windmills were a portent of what was to follow and opened up an era of experimental and laboratory work for the mechanical engineer comparable with the laboratory experiments of the scientist. After what had gone before it is not surprising that in this period the centrifugal pump was developed almost to its present stage and this was also true of some types of water turbine such as the Francis and the Thomson. On the other hand, the windmill had begun to lose its importance for grinding flour from the time when the Albion Mills in London were equipped with steam power in 1784 which revolutionized the flour industry. Most significant for the future were the balloon trials of Montgolfier and Blanchard and the glider experiments and analysis of the requirements for flying heavier-than-air machines made by Sir George Cayley.

As the main constructional material for machines, wood had at last been displaced by iron and the credit for this should undoubtedly go to Smeaton. Iron became the natural material for the engineer to work in, first cast iron, then wrought iron, and finally, steel. Shortly after the period began in 1769 came the first iron (cast iron) bridge. The early railway bridges were built of wrought iron, whereas steel later supplanted this as the material of choice, and a similar sequence was followed in the materials used for the rails. England continued to retain an astonishing lead in the production of iron, due to the successive rapid adoption of Cort's puddling process for making wrought iron, Neilson's hot-blast stove for the blast furnace, and later Bessemer's converter for making steel. A similar lead was achieved and maintained in the working of metals, through the work of Bramah, Maudslay, Whitworth, Roberts, Nasmyth and others, who established between them standards of accuracy of workmanship previously unknown.

Much of the state of mechanical engineering at this time is revealed in Professor Charles Babbage's book *Economy of Manufactures*, published in 1832. He wrote the book after visiting factories in Britain and on the Continent, to see what facilities were available for making his famous 'difference engine' or calculating machine. He found that one result of the Industrial Revolution and the use of the steam engine in England had been a large reduction in the cost of manufactured goods, so that between 1818 and 1830 English pig iron had fallen in price from £6 7s. 6d. to £4 10s. 0d. per ton, and bar iron from £10 10s. 0d. to £6 0s. 0d. per ton. Also, large sizes of plate glass in 1832 were less than half the price of forty years before, though this was not so either in Paris or Berlin. As a rule all manufactured products were very much cheaper in England than on the Continent. His book also contains a detailed analysis of the cost of the individual processes involved in the making of pins, perhaps the first example of detailed cost accounting in a factory.

Engineers of today are frequently surprised to learn that so many of the machine tools we use now were invented so long ago, such as the boring mill, the planer and the shaper, in England, and the milling machine, the turret lathe and nearly all the various types of gear-cutting and grinding machines in the United States. The latter country, though not yet a serious challenger of England's supremacy in mechanical engineering, had begun to show that peculiar genius in production engineering that is still in evidence. She had also demonstrated the ability to devise ways of mass production for revolvers, sewing machines and clocks, and, more regrettably, had shown a reckless daring in building unsafe types of high-pressure steam engines and boilers, sometimes with disastrous consequences, as in the paddle boats of the Western rivers.

In the field of machines, it was the application of power to the spinning and weaving of textiles that had the greatest influence in creating the factory system. The movement from a cottage industry into the factory had begun when the power from water wheels was used for fulling, but the invention of Arkwright's water frame, Hargreaves's spinning jenny, Crompton's mule, and finally the power loom, meant that the operations of producing cloth from natural fibres must inevitably be concentrated in the factory, where one power unit could supply a large number of machines at the same time. Thus the desultory habits of work of the mediaeval worker had to give way to the discipline of regular hours of work. Though some attempts to start yarn and cloth factories in America at this time failed,* one of the greatest mechanical successes in an allied sphere took place there when first Elias Howe and later Isaac Singer invented a domestic sewing machine. The Singer Machine of 1851 had most of the features of its modern counterpart, and became so popular and was required in such large numbers that its manufacture provided opportunities for the American tool builders to develop special machines for making components with high accuracy and low cost.

In marked contrast to the ceaseless activity of the British and American practical men were the scholarly studies of the engineering scientists on the Continent, who were developing the science of strength of materials. Nothing comparable with Navier's book on this subject then existed in the English language, and we owe to men such as him our ability to calculate the strength of a beam or a column or the dimensions of a cylinder or shell needed to withstand a given internal pressure. The continental engineers were able to make such calculations with confidence and were the first to verify them by experiment. There were very few engineering scientists in England or America, a fact only realized by one or two perspicacious individuals, this being in marked contrast to the situation in France, where a succession of brilliant men, well trained at the École Polytechnique, were laying the foundations of the science of strength of materials and developing hydraulic machinery and the theory of hydraulics. It is a singular but undeniable fact that in England immense commercial value emerged from the work of comparatively uneducated or at best self-educated men such as James Watt, George Stephenson, Richard Trevithick and Henry Maudslay and that in contrast the work of the theorists, on the Continent and elsewhere, brought them at that time little financial reward or recognition, although their contributions were more fundamental to the basis of engineering science. Throughout this period the training of mechanical engineers in Britain and in the United States lagged far

* Reference 2 in Chapter VII.

behind that of the Continent of Europe where, particularly in France, courses of training based on the study of mathematics, physics and chemistry and their application to the design of machinery, were already far advanced.

REFERENCES

(for abbreviations see List of Acknowledgments)

1. Mumford, L., *Technics and Civilisation*. G. Routledge and Sons, London, 1934.
2. Catalogue of the Great Exhibition of 1851.
3. Timoshenko, S. P., *History of Strength of Materials*. McGraw Hill, London, 1953.
4. Rouse, H., and Ince, S., *History of Hydraulics*. Iowa Inst. Hyd. Res., State University, Iowa, 1957.
5. Hodgson, J. E., 'Sir G. Cayley as a Pioneer of Aeronautics.' *T.N.S.*, vol. iii, 1922, p. 69.
6. Collar, A. R., 'Great Engineers: Sir George Cayley', *The Times Educational Supplement*, 1956.
7. Hulme, E. W., 'The Career of Benjamin Huntsman', *T.N.S.*, vol. xxiv, 1943–5, p. 37.
8. Raistrick, A., *Dynasty of Ironfounders: The Darbys of Coalbrookdale*. Longmans Green, London, 1953.
9. Rhodin, J. G. A., 'Christofer Polhem', *T.N.S.*, vol. vii, 1926–7, p. 17.
10. Petree, J. F., 'Maudslay Sons and Field as General Engineers', *T.N.S.*, vol. xv, 1935–6, p. 39.
11. Dickinson, H. W., 'Richard Roberts, his Life and Inventions', *T.N.S.*, vol. xxv, 1945–7, p. 123.
12. Roe, J. W., *English and American Toolbuilders*. Yale U.P., 1916.
13. Willis, R., *Principles of Mechanism*. J. W. Parker, London, 1841.
14. Harris, L. E., 'Some factors in the early development of the centrifugal pump', *T.N.S.*, vol. xxviii, 1951–2, p. 187.
15. Hunter, L. C., *Steamboats on the Western Rivers*. Harvard U.P., 1949.

BIBLIOGRAPHY

Babbage, C., *The Economy of Machinery and Manufactures.* Charles Knight, London, 1832.

Buchanan, R., *A Treatise on Mill Work.* London, 1841.

Church, W. C., *The Life of John Ericsson,* 2 vols. Sampson Lowe and Marston, London, 1892.

Dickinson, H. W., *A Short History of the Steam Engine.* C.U.P., 1938.

Dickinson, H. W., 'Henry Cort', *T.N.S.,* vol. xxi, 1940–1, p. 31.

Dickinson, H. W., 'Joseph Bramah and his inventions', *T.N.S.,* vol. xxii, 1941–2, p. 169.

Dickinson, H. W., and Jenkins, R., *James Watt and the Steam Engine.* O.U.P., 1927.

Dickinson, H. W., and Titley, A., *Richard Trevithick, the Engineer and the Man.* C.U.P., 1934.

Ewbank, T., *Hydraulic Machines.* Tilt and Bogue, London, 1841.

Farey, J. A., *Treatise on the Steam Engine: historical, practical and descriptive.* London, 1827.

Hunter, L. C., *Steamboats on the Western Rivers.* Harvard U.P., 1949.

Mason, W. W., 'Trevithick's First Rail Locomotive', *T.N.S.,* vol. xii, 1932.

Pendred, L. St. L., 'Trevithick', *T.N.S.,* vol. i, 1920, p. 34.

Raistrick, A., *Dynasty of Ironfounders: The Darbys of Coalbrookdale.* Longmans Green, London, 1953.

Robinson, T. R., 'Centenary of the Great Clock of Westminster', *Horological Journal,* vol. 101, 1959.

Roe, J. W., *English and American Toolbuilders.* Yale U.P. 1916.

Rolt, L. T. C., *Isambard Kingdom Brunel.* Longmans Green, London, 1957.

Rolt, L. T. C., *George and Robert Stephenson.* Longmans Green, London, 1960.

Rouse, H., and Ince, S., *History of Hydraulics,* Iowa Inst. Hyd. Res., State University, Iowa, 1957.

Singer, C. (editor), *A History of Technology,* vol. iv. O.U.P., 1958.

Smeaton, J., *Reports of the late John Smeaton,* 4 vols. London, 1812–14.

Smiles, S., *Lives of the Engineers.* John Murray, London, 1878.

Smiles, S. (editor), *Autobiography of James Nasmyth*. John Murray London, 1891.

Symonds, R. W., *A Book of English Clocks*. King Penguin, London, 1947.

Thomas, J., *The Story of George Stephenson*. O.U.P., 1952.

Thurston, R. H., *A History of the Growth of the Steam Engine*. C. Kegan and Co., London, 1878.

Timoshenko, S. P., *History of the Strength of Materials*. McGraw Hill, London, 1953.

Titley, A., 'Richard Trevithick and his winding engine', *T.N.S.*, vol. x, 1930, p. 55.

Wailes, R., *The English Windmill*. Routledge and Kegan Paul, London, 1954.

Willis, R., *Principles of Mechanism*. John W. Parker, London, 1841.

Wilson, P. N., 'The Water Wheels of John Smeaton', *T.N.S.*, vol. xxx, 1955–7, p. 25.

Woodbury, R. S., *History of the Grinding Machine*. M.I.T.P., 1959.

Woodbury, R. S., *History of the Gear Cutting Machine*. M.I.T.P., 1958.

Woodbury, R. S., *History of the Milling Machine*. M.I.T.P., 1960.

CHAPTER VII

The Age of Steam Power, 1850, 1900

Between 1850 and 1900 there was a great upsurge of interest in the training and education of engineers in all the industrialized countries of the world. In Britain the movement was led by the Prince Consort, who envisaged an engineering college for the whole of the British Empire—in South Kensington—together with its industrial museum and science library as a permanent memorial of the success of the Great Exhibition of 1851. In the City and Guilds College, professorships were established in many branches of engineering and were soon occupied by such distinguished men as Unwin, Ayrton, Perry and Sylvanus P. Thompson, who, with others, shaped for engineers at British universities the courses of instruction which endured with little change for more than fifty years. Other institutions also contributed in no small measure towards the same goal. Universities or university colleges were established in Manchester, Liverpool, Leeds, Bristol, Newcastle, Nottingham, Sheffield and Birmingham, all with engineering courses and at last, in 1875, an engineering department was established at the University of Cambridge. Among the professors of engineering of this time the names of Osborne Reynolds, Alexander Kennedy and Goodman are still familiar, but the brightest light of them all was W. Macquorn Rankine of Glasgow, whose textbooks on heat engines and applied mechanics set new standards of mental discipline for students of mechanical engineering.

In Germany something similar was done by Professor Reuleaux, whose textbook on machines remains a classic to this day. Technical colleges and universities were enlarged all over Germany and engineering courses were established in institutes in most of the smaller countries in Europe—the most famous for engineering being the Federal Polytechnic Institute in Zurich. Even less-developed countries such as Japan had established courses for engineers before 1900, as had universities in Australia, South Africa and Canada, but it was in the United States of America that the most spectacular developments took place, and there the rates of industrial growth and of the establishment of higher technical education began to outpace the rest of the world, so that by the turn of the century the engineering teaching at many American universities could bear comparison with that in any of the older countries and a much larger proportion of the population was being taught. Establishments such as the Massachusetts Institute of Technology, and the Universities of Harvard, Yale and Princeton, to mention only a few, were making a wholesale contribution to the training of engineers.

The movement in England was not confined to higher technology. For those less well educated, mechanics institutes were established, where workmen could attend classes in the evenings to 'improve themselves' as it was called. From such humble beginnings, many of the best technical colleges in Britain were developed; in 1876 the City and Guilds of London Institute was established and proceeded to conduct examinations in technical subjects for workers studying in the evening institutes.

It was still the general belief in Britain—though not in all other countries—that the mechanical engineer's training should be primarily in the workshop and that the theoretical knowledge he gained at the university was of very minor importance. An exception was Sir Joseph Whitworth, who established in 1868 thirty 'Whitworth' Scholarships for young engineers who were combining theory with practice, in training to be mechanical engineers.

Britain provided a succession of scientists whose ideas and analytical work in mathematics and physics supplied a basis for important developments that were to be made by later generations of mechanical engineers. Most of these men received their scientific training in the hard school of the mathematical tripos at Cambridge. The name of Stokes is known to engineers by the law which bears his name that relates to the rate of fall of a solid sphere in a viscous fluid, but he also made contributions to the theory of elasticity and established theorems relating to the vibration of elastic bodies. William Thomson, later Lord Kelvin, who was Professor of Natural Philosophy at Glasgow University from 1846 until 1899, was the leading British physicist of the period, making the most important discoveries in thermodynamics, electricity and the properties of matter. Michael Faraday of London did work on electromagnetic induction which led to the birth of electrical engineering. James Prescott Joule of Manchester established the mechanical equivalent of heat and by his work confirmed the law of the conservation of energy. James Clark Maxwell in his short life—he died at 48—established a general theory of three-dimensional stress systems, gave engineers a simple and accurate method for determining the stresses in a framework, contributed to the kinetic theory of gases, and published a famous memoir on electricity.

It is not generally appreciated that before Maxwell went to Cambridge at the age of nineteen (in 1850) he had completely developed the technique of photo-elastic stress analysis in two dimensions using different kinds of glass and other transparent materials. In the last years of his life he set up the original Cavendish laboratory for experimental work in physics at the University of Cambridge, and was succeeded as professor of experimental physics by J. W. Strutt, third Lord Rayleigh, whose fundamental work on vibration and elasticity is still used by mechanical engineers concerned with these problems. Rayleigh also made important contributions to hydro-dynamics, particularly in cavitation, the determination of wave profiles and the instability of jets, and was among the first to popularize the principle of dynamic similarity. The most comprehensive writer on hydrodynamics at this time was Horace Lamb, whose textbook on the subject, first published in 1879, is still a standard book of reference at the present day. Lamb wrote many other papers and books on mathematics and physics but none of these has stood the test of time so well as his *Hydrodynamics*.

Another famous textbook of this period that has stood the test of time is *The mathematical theory of elasticity* by A. E. H. Love, which first appeared in 1892. This book has been translated into other languages and used as a reference book of the subject all over the world.

It is not easy to find eight physical scientists of equal eminence in any other country during this period. On the Continent, to be sure, there was Clausius, who formulated the second law of thermo-dynamics in 1850 and died in 1888, Helmholtz the German physicist who made such substantial contributions to the analysis of fluid motion, and Neumann who considered theoretically the internal residual stresses in materials after plastic deformation and elaborated the theory and technique of photo-elastic stress analysis.

One brilliant star that could compare with any of these shone brightly from across the Atlantic at this time. Thomas Alva Edison was both a scientist and an engineer. Without formal education he was able by sheer genius to solve scientific and engineering problems that baffled other men.

MATERIALS

The outstanding development of this period was the making of mild steel, both by the Bessemer converter and in the open hearth furnace, and the steps taken by mechanical engineers to satisfy themselves that steel so made was a suitable material of construction for machines and structures. In Britain, David Kirkaldy established a materials testing laboratory in London in 1865 and on the Continent materials testing laboratories were established in the engineering schools at Münich and Berlin, 1871, Vienna, 1873 and Zurich and Stuttgart in 1879. Other laboratories were established in Sweden, Russia and elsewhere so that when the first international congress on the testing of materials was called in 1884 no fewer than seventy-nine research workers from different countries attended. The leader

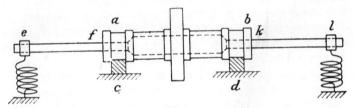

Fig. 186. Wöhler's testing machine.

in this was Professor Bauschinger of Münich who had installed a 100-ton testing machine and had devised a mirror type of extensometer capable of measuring unit elongations to an accuracy of one millionth part. Bauschinger and Kirkaldy both recognized the importance of the shape and size of the test specimen and the necessity for using geometrically similar test pieces. Accordingly one of the first tasks of the international conference was to draw up specifications for the testing of materials.

The importance of the fatigue strength of metals was brought to the fore by experience on the railways, where fractures of railway axles and joints were a common occurrence that cried out for a remedy. The leading authority on this matter was A. Wöhler, a railway engineer of Hanover who, before 1858, devised his now famous fatigue-testing machine to study the causes of this phenomenon. In Wöhler's machine (Fig. 186) two cantilever specimens *ef* and *kl* are rotated at one end by the central pulley in the figure while their extremities are loaded by the springs shown operating through ring bearings. The stress in the specimens is determined by adjustment of the springs. It will be seen from the figure that during each revolution of the specimen the stress on the outer fibres of the material passes through a complete cycle from tension to compression. Machines of this type are still in use for testing the fatigue strength of metals and have yielded an immense amount of information from which data for safe design have been determined.

In the field of mathematical analysis in strength of materials, structures and the theory of elasticity, many engineers on the Continent made valuable additions to knowledge and the reader interested should consult Timoshenko's *History of the Strength of Materials*. Most important to the mechanical engineer were the contributions of Saint Venant, Winkler, Castigliano, Mohr and Hertz.

Saint Venant was the outstanding authority during this period (he died in 1886 at the age of 89) and occupied himself to the end of his life with the analytical problems that he so much enjoyed. It was he who proposed the semi-inverse method for the solution of problems in torsion in which some features of the displacements and forces are assumed and the others are found by satisfying the elastic equations. He went on to apply the method to the bending of a cantilever and then considered problems in combined bending and torsion. Saint Venant recommended that in designing beams the maximum strain theory should be used and a great many continental designers followed this advice. Also, he was interested in the effect of longitudinal impact on bars and showed that the maximum stress at impact occurs at the fixed end of the bar. Saint Venant was also the first to derive fundamental equations for the theory of plasticity and by making certain assumptions he solved some of the simple cases of plastic deformation.

E. Winkler in 1858 tackled the intractable problem of the bending of curved bars, as in hooks, rings and chain links, and produced a solution showing that the conventional beam theory did not apply if the dimensions of the bar were large compared with the radius. Winkler also investigated the strength of tubes subjected to internal and external pressure and the stresses in rotating discs and flywheels. He was also the first to investigate the bending of a beam on an elastic foundation and to show its application to the stresses in rail tracks. He wrote a textbook on strength of materials that was used for many years by German engineers.

Castigliano, an Italian engineer who died at the early age of 37, produced his famous theorem of least work for the analysis of structures in his degree thesis in 1873. Two years later the theorem was extended to apply to elastic solids of any shape. Castigliano was one of the earliest workers to use the concept of strain energy in structural analysis and his principle of least work has been used ever since in the analysis of structures containing redundant members.

The name of Otto Möhr is known to engineering students for his graphical representation of the stress at a point, that is so useful in teaching. He showed in 1868 that for a two-dimensional system the normal and shearing forces of the stress on a plane may be defined by the co-ordinates of points on a circle whose diameter is equal to the difference of the principle stresses. Similarly he showed that for the three-dimensional case the stress components can be determined from three circles of which one circumscribes the other two. Möhr also made useful studies of the bending of plates and shells, elastic stability in buckling problems and the vibration of bars and plates.

Heinrich Hertz, who began working with the physicist Helmholz in Berlin in 1878, concerned himself with some unusual problems in elasticity. He studied the compression of elastic bodies during impact and calculated the duration of the impact of two elastic spheres; he then determined the stresses in the spheres during impact, an exercise that proved useful to later generations of engineers when they came to design ball bearings and wanted to know the maximum loading that they could withstand. Hertz also investigated the stresses in rollers; he was one of the first workers to interest himself in the hardness of materials and proposed a test for hardness that has never been accepted—depending upon the load required to produce the first crack.

Between 1850 and 1900 the Iron and Steel Industry was revolutionized by the coming of cheap steel. Two distinctly different methods were discovered for achieving this—first the Bessemer process, announced by Sir Henry Bessemer in 1856, and a few years later the Siemens Martin open hearth process which was to become the more popular of the two before the end of the century.[1]

Steel-making is primarily a chemical process and the early development of these two processes suffered from lack of sufficient knowledge of its chemistry, which was little understood at the time, particularly by the ironmasters and engineers responsible for its development.[2] The leading American steel magnate of this period, Andrew Carnegie, owed some of his success to his being among the first to employ a chemist in a steelworks.

291

The essential feature of the Bessemer process is that by blowing air through liquid pig iron the metal can be decarburised to the degree required to produce steel. No fuel needs to be added, for the carbon in the pig iron acts as the fuel, burning in the air blast to produce the higher temperature needed just where it is required—in the

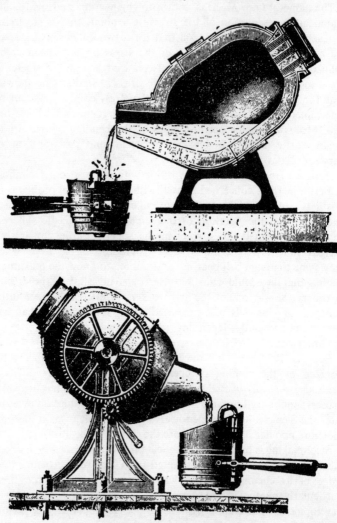

Fig. 187. Bessemer converter.

molten metal itself. The process was independently invented by Bessemer in England and by William Kelly in the U.S.A. Kelly was the first inventor and was granted a U.S. patent, but became bankrupt the same year and was unable to exploit his invention. Bessemer who (like Kelly) began his experiments with a stationary furnace, later made his furnace tiltable so that the metal should not choke the tuyère or blowing nozzle while it was being poured, as shown in Fig. 187. Before blowing, the converter, shown in outside view in the lower part of the figure, was rotated anti-clockwise through an angle of about 135 degrees. During blowing, the air blast at 10–15 lb./sq. in. had sufficient pressure to force its way up through the molten metal. The whole cycle occupied less than 30 minutes so it was a rapid method of making steel.

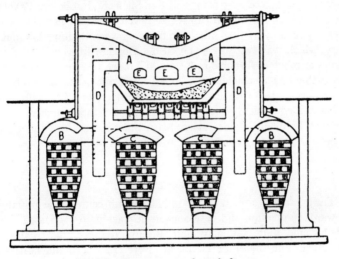

Fig. 188. Siemens open-hearth furnace.

The Siemens open hearth process (Fig. 188), introduced in England in the 1860s, came to be preferred by the largest steel producers in Britain and the United States after Pierre Martin, a Frenchman, had shown in 1864 that steel could be made in such a furnace by adding a large proportion of scrap metal to the pig iron. The re-use of scrap was attractive on economic grounds and the process was also more suited to some of the available ores than the Bessemer process and was easier to control to obtain the quality of steel desired. Heat for the process was supplied from outside—from coke oven gas, or gas producers or by oil or natural gas. The chequer brickwork regenerators marked B and C on the figure were below the furnace hearth E where the metal was heated by direct contact with a long horizontal flame of gas burning with preheated air supplied through the regenerators. After combustion in the furnace the hot gases passed out through the chequerwork of the regenerators which were heated up in readiness to be cooled down again by incoming air after the periodical changeover, when the flow of air and gas was reversed. By this means a temperature at the hearth of about 1,650° C. could be maintained.

That the commercial production of cheap steel by these processes was established in Britain ahead of other countries is confirmed from the report that by 1863 steel plates twelve inches thick and weighing twenty tons apiece were being rolled in Sheffield, but before 1900 the lead had been lost to the U.S.A., and even Germany had surpassed her. By then the largest American furnaces were smelting 800 tons a day, which was twice as much as the German furnaces and three times as much as furnaces in Britain or France.[2] The greater productivity that had been achieved in America was due to the process being more mechanized with larger and better furnaces operated with more advanced equipment, better refractories, better thermometers, more powerful blowing engines, steam blast, water jackets for cooling and mechanical loading machines.[2]

TOOLS

During the second half of the nineteenth century the initiative towards innovation and development in machine tools passed completely and quite suddenly from Britain to the United States of America. The major developments were the automatic lathe for making screws and small components, the twist drill and the universal milling machine for milling the flutes, the Brown and Sharp milling cutters, the hobbing process for making gears and the synthetic grinding wheels which made the grinding process a method of production superior to any other for many purposes. All these improvements were developed in the United States, and the British workshops, instead of adopting them readily, stuck obstinately to their own methods—planing and shaping instead of milling, and hand scraping in preference to surface grinding. Some machine tool developments took place outside the United States at this time but they were unimportant by comparison with those mentioned above.

The preoccupation of British and Continental inventors seems to have turned from machine tools to heavy equipment for the production of metals, as in the Bessemer converter (Fig. 187), the Siemens open-hearth furnace (Fig. 188), the three-high rolling mill for the hot rolling of steel sheets and sections (Fig. 190), and the Mannesmann process for making seamless metal tubing (Fig. 189).

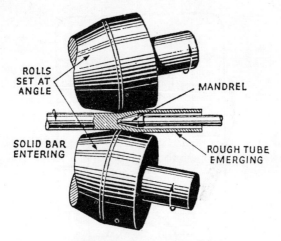

Fig. 189. The Mannesmann Process

The process announced by the brothers M. and R. Mannesmann in 1885 involved an entirely new principle in metal forming that had been earlier conceived by R. Mannesmann at Remscheid in Germany in 1860. A hot bar from the furnace was fed between the two rolls (shown in Fig. 189) which, owing to their geometrical configuration, drew the bar forward and at the same time produced tensions in the hot metal that caused it to tear apart at the centre. A stationary pointed mandrel induced the ingot to open out and form a tube. The tubes were later forged to size (on mandrels) in rolling mills with grooved rolls that varied in diameter like cams. This latter process, which is of necessity intermittent, is called pilgering. The first Mannesmann plant and pilger mills were installed in Swansea in 1887 and were able to pierce billets up to ten inches diameter. The process—still widely used—is a most unusual one and caused a sensation when it was announced. It was the invention which most impressed the American inventor T. A. Edison during his visit to the World Exhibition in Chicago in 1893.

The idea of the three-high mill (Fig. 190) for rolling metal was more than a century old before it was first introduced into iron-works at Motala in Sweden in 1856 and at Birmingham, England in 1862. Its inventor, Christopher Polheim (*see* Chapter V), had realized the value of being able to pass the metal quickly back and forth without having to reverse the rolls. Another idea from the previous

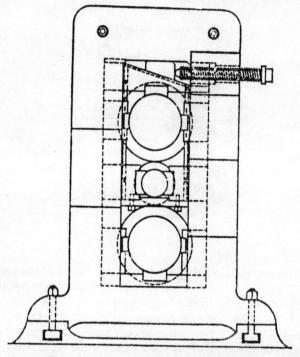

Fig. 190. Lauthes's three-high mill, 1862.

century, the continuous mill, had been patented by William Hazeldine in 1798 but was not used until it was re-invented by George Bedson of Manchester in 1862. Here the metal was fed successively into a series of roller stands placed in line so that its size was reduced at a single heat. Continuous wire mills for the cold rolling of miles of copper wire for the electric telegraph were also introduced at this time.

The British took a leading part in improving machinery for forging metals, first by additions to the steam hammer invented by Nasmyth in 1838; this was made double-acting in 1850. Some years later pneumatic hammers were introduced in which a small air compressor was attached to the back of the machine. Drop hammers also began to be used for light work. The hydraulic press was applied to metal-working, particularly in England where Armstrong was promoting the use of hydraulic power for many purposes. Hydraulic presses were employed for hot forging, extruding and for the cold drawing of thin metal in dies, as in making metal holloware. Holloware was also made by spinning in a lathe, the metal sheet being rolled over a wooden former of the shape required or sometimes rolled over by the spinner 'against the wind' i.e. without any former behind it. The making of light articles of spun brass and copper became very popular as in the hands of a sufficiently skilled man it could be done very quickly indeed and owing to the heat produced the metal became annealed and troubles due to work-hardening did not arise.

Machinery for making sand moulds for iron castings was used in the 1850s. One of the earliest machines was described by R. Jobson.[3] Similar to a modern simple moulding machine, it consisted of a table on trunnions, enabling the table to be turned over with the mould and pattern on it to lessen the labour of lifting the mould. Ramming of the sand was done by hand and it was claimed that with the machine an operator could make a mould for two railway chairs (for rails) every minute. There were arrangements for casting near the machine so that the moulds could be filled as fast as they were made.

In 1853 the Tangye brothers established works in Birmingham, England, to make hydraulic jacks and other hydraulic equipment. In 1858 they supplied a large number of small jacks for the launching of I. K. Brunel's giant steamship *Great Eastern* in the Thames. Thereafter they expanded into the manufacture of hydraulic presses, portable hydraulic shearing and riveting machines, portable hydraulic punching machines and in 1863 a large hydraulic shearing machine for Russia was made weighing 24 tons and able to exert a force of 1,000 tons with a 16-in. diameter ram.[4] The Tangye brothers also made screw jacks and it is reported that in 1863 Joseph Tangye built a machine that cut the threads of six screws simultaneously. A few years later he constructed one of the first thread-rolling machines for forming the screw threads by rolling instead of cutting.

Some years earlier, in 1860, an interesting drilling machine with hydraulic feed was described by J. Cochrane.[5] It was for drilling the wrought-iron side plates ($\frac{5}{8}$ in. thick) of the Hungerford Bridge over the River Thames at Charing Cross. This machine (Fig. 191) was able to drill eighty 1-in. diameter holes simultaneously without marking off and with great accuracy of pitch. The table on which the plate was raised to the drills was counterbalanced and operated hydraulically, the pressure while drilling being sustained from an accumulator providing a total load of 20 tons, or $\frac{1}{4}$ ton per drill.

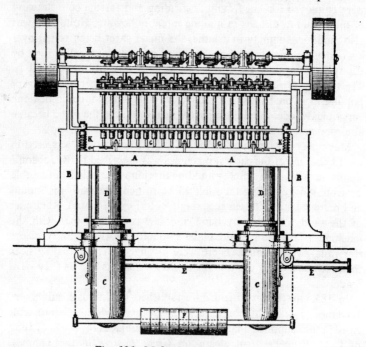

Fig. 191. Multiple drilling machine.

Flat arrow-headed drills were used, running at 50 rev./min.; the time of drilling was 15 minutes per plate and the life of the drills between re-grinds about ten hours.

Nasmyth devised a slot drilling and grooving machine that had a certain vogue during the 1860s. Such slots were required for cotters on piston rods and for cutting key-ways; the secret of the success of his machine was due to the cutter, which was a flat arrow-head drill with the centre removed, and the extremities ground away like a face and edge milling cutter. This machine tool, like Nasmyth's nut-milling machine of 1830, had some resemblance to a vertical milling machine except that it had intermittent feed for the work. In the slot miller the work was traversed to and fro by a crank and at the end of each stroke the cutter or drill was fed downwards. Nasmyth, who built a large number of these machines, claimed that they were self-acting and that a boy could superintend two of them quite easily.[6]

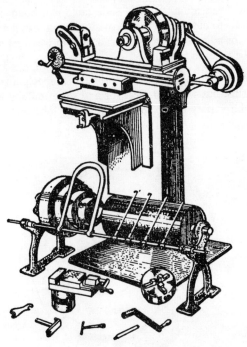

Fig. 192. First universal milling machine.

The twist drill (*see* Fig. 201) appeared in the U.S.A. in 1860. Its effect on the practise of drilling and the design of drilling machines was considerable, for it was soon found that higher speeds for drilling, and higher rates of feed for the drills could be used if the drilling machine was sufficiently strong and stiff to permit it. The spiral flutes on the first twist drills were made by hand filing which was slow and costly, so in 1861 F. W. Howe suggested to J. R. Brown of the Brown and Sharpe Company in the United States that a special milling machine should be designed for cutting the spiral flutes. Brown did this in 1861 and thereby produced the first universal milling machine (Fig. 192). The new features which made this machine the true ancestor of the modern universal milling machine have been described by Woodbury[7] as:

'(1) A work table carried in swivelling saddle mounted in a clamp bed which in turn has a transverse movement on the vertically adjustable knee
(2) A spiral head mounted on a slide table and connected to the feed screw to cut spirals with provision for indexing and
(3) Means for adjusting the spiral head, so that when set at an angle it can be indexed.'

This invention not only solved the problem of machining the flutes on twist drills, but was also ideal for any spiral milling work, gear cutting and a variety of the special machining jobs that arise in the workshop, since its flexibility was remarkable.

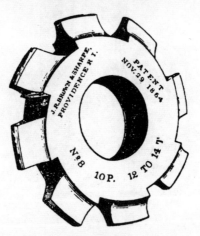

Fig. 193. Brown's milling cutter.

The rapid adoption of the milling machine in the United States would not have occurred without Brown's invention in 1864 of the formed cutter (Fig. 193) which could be ground on its face without changing its shape. A great variety of other cutters for different types of milling soon made their appearance.

The vertical milling machine (Fig. 194) was in use in 1860 not only in the United States but also in Britain, France and elsewhere in Europe. Improvements were being made throughout this period, such as power feed for the traverse and crossfeeds, the swivelling table, and reversing gear, so that after the turn of the century the foundations had already been laid for the development—mainly in Europe—of the jig boring machine.

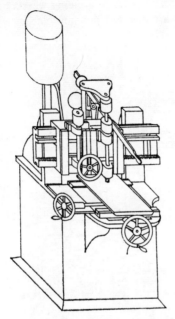

*Fig. 194. Vertical milling
machine.*

Professor Woodbury has recently reviewed in considerable detail the great activity that took place in developing new types of machines for gear cutting.[8] Most of this occurred in the United States and culminated in two outstanding achievements during the last decade of the century; these were the Fellows gear-shaping process and the hobbing process for making gears of all kinds. E. R. Fellows took out patents in 1897 for a gear-shaping machine and a cutter generator for making the cutters used in the gear shaper. His work merited the award two years later of the John Scott Medal by the Franklin Institute. The Fellows process worked on the principle of two gears rolling together in mesh and so is described as a moulding generating method because during manufacture the gear blank being cut represents one of the gears and rotates but does not reciprocate while the gear-shaped cutter, with its cutting edges relieved, both rotates and reciprocates. Fellows was the first to use a complete gear wheel instead of a rack tooth as the generator (or cutter) and clearly an essential part of the process was the means adopted for sharpening the cutter. By this time suitable grinding wheels were available and Fellows designed a cutter grinding machine in which the involute

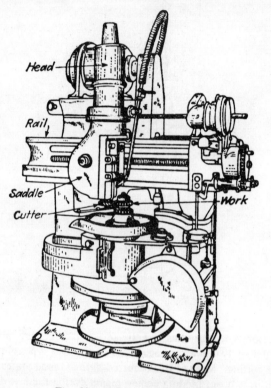

Fig. 195. Fellows's gear-shaper.

shape of the cutter was generated afresh while grinding by making the grinding wheel represent one side of the tooth of an involute rack while the cutter being ground rolled over it. The required movement of translation was obtained by two flexible steel bands moved by the cutter as it was rotated during grinding. Fig. 195 shows the outline of a modern Fellows gear-shaping machine for cutting spur gears.

Hobbing is a continuous milling process in which a milling cutter shaped like a worm—the hob—rotates in fixed relationship to the gear blank being cut, so that a generating action occurs, for when a single thread hob is used the hob makes one revolution while the gear being cut moves about its axis the distance of one tooth and one space. The hobbing process conceived in 1856 by Christian

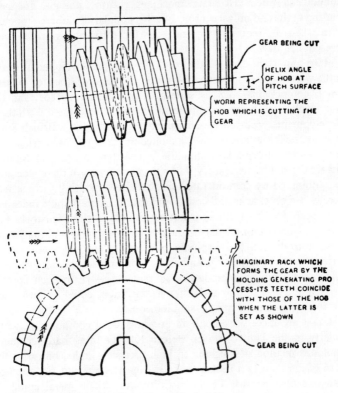

GEAR BEING CUT

HELIX ANGLE OF HOB AT PITCH SURFACE

WORM REPRESENTING THE HOB WHICH IS CUTTING THE GEAR

IMAGINARY RACK WHICH FORMS THE GEAR BY THE MOLDING GENERATING PROCESS-ITS TEETH COINCIDE WITH THOSE OF THE HOB WHEN THE LATTER IS SET AS SHOWN

GEAR BEING CUT

Fig. 196. Principle of the hobbing process.

Schiele became a practical one for production work as soon as involute-shaped gear teeth superseded the cycloidal type in the 1880s, since the involute hob, like the involute rack, has straight sides (for the worm is a form of continuous rack) so that to make a hob from a worm all one has to do is to gash some teeth in the worm so that it will cut the blank as it is rotated. The principle of the hobbing process of forming spur gears is shown in Fig. 196. It will be seen that there is a rolling action of the gear being cut in relation to the hob so that an involute contour is produced on the gear teeth. In this respect the hobbing process differs from the process of milling gear teeth, where each tooth is formed in turn to its full depth and has a contour determined only by that of the cutter. In milling spur gears, the blank is stationary while each tooth is formed and is then moved round one pitch. Hobbing machines became popular for gear cutting at the end of last century but this popularity gathered momentum when the requirements for gears for automobiles arose in the present century.

Two important changes that took place in the development of the lathe were the design of special-purpose automatic lathes for making small components such as screws, which started in 1861, and the introduction of the multi-spindle automatic lathe in 1895. Both these developments occurred in the United States, though a British automatic lathe for making screws, patented in 1879 by C. W. Parker, was built and multi-spindle automatic lathes were on sale in Sweden before the end of the century. The automatic action of these machines was obtained by cam-operated mechanisms operating turrets, feed screws, toggle chucks and collets and for the single spindle machines bar stock was fed through a hollow spindle in the headstock. For feeding small components to the machines, hoppers and magazines were introduced. Multi-spindle machines were provided with a carrier on the main spindle supporting four or five secondary spindles which were moved periodically so that the work in each spindle arrived at a new station and different operations could be performed on the work in each spindle simultaneously.

Great advances were made in grinding metal components, particularly in the United States. Before 1900 grinding had become a standard method of finishing to high precision limits, but after the turn of the century it became acknowledged as a production method valuable also because of the rapidity with which metal could be removed in the process. The application of grinding to engineering work has always been dependent upon the form and quality of grinding material available as well as upon the type of machine used. Artificially made grinding wheels, of a sort, were being used in England, France, Germany and the United States before 1850. They were made of emery—aluminium oxide, iron oxide and silica bonded together to form a wheel, the bond being made of glue or baked clay, vulcanized rubber or vitrified silicate. Professor Woodbury[9] has recently reviewed the sequence of events that led to the rediscovery by E. A. Acheson in 1891 of how to make Carborundum, the name he gave to artificially produced silicon carbide. As it was made in an electric furnace, requiring cheap power, works were established at Niagara Falls, where Carborundum grinding wheels were being made commercially by 1896. Meanwhile Brown and Sharpe had produced (in 1876) their universal grinder, a machine very similar in essential respects to the modern universal grinder, except that it had belt drives. It was a robust precision machine capable of extremely

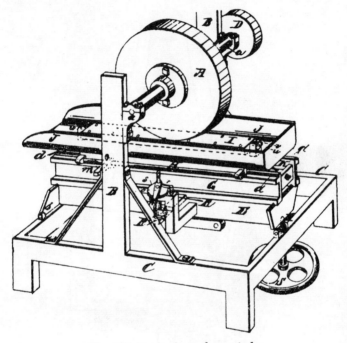

Fig. 197. Darling's surface-grinder.

accurate work, and was the first milestone in the history of grinding for precision manufacture of metal components. Much earlier, in 1853, S. Darling had patented a surface grinder, Fig. 197, which had a wheel mounted on horizontal bearings, vee-type slideways for the table, and below the slideway supports there was a saddle, also on vee-guides, for providing the crossfeed. Vertical control of the work was provided by the screw and handwheel shown in the figure. This machine was used for many years by Brown and Sharpe who had been joined by Darling in 1866.

Two other developments in grinding stand out as particularly important in this period. The first is J. M. Poole's precision roll-grinding machine (Fig. 198). This introduced a new principle in that the roll being ground was itself used to determine the straightness and

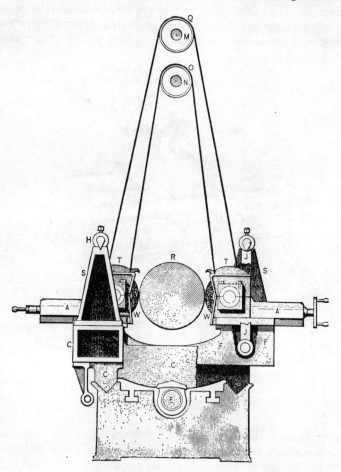

Fig. 198. Poole's roll-grinder.

uniformity of the finished work. This was because each of the two grinding wheels was supported in a swinging frame balanced on knife edges (H and J in the figure) so that the wheels could swing in a plane at right angles to the roll. The wheels were adjusted by the cross slides A, so that the distance between them at the bottom of the swing was equal to the roll diameter required. If the roll was oversize the wheels would be pressed against the roll by a gravitational component until the finished size was reached, when the side pressure would be insufficient for grinding to continue. Poole's machine was designed for finishing heavy rolls for paper mills. The accuracy of the work was tested in 1886 by Professor Sweet of Cornell University and found to be within 0·000025 in. on the diameter.[10]

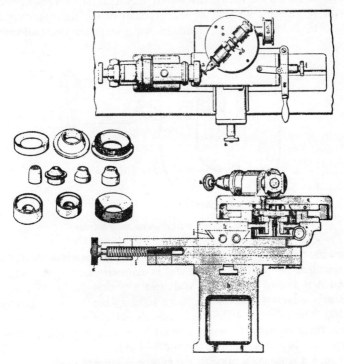

Fig. 199. Grinder for cup-and-cone bicycle ball-bearing.

The other remarkable development in grinding was the machine produced by Pratt and Whitney in the United States in 1899 for grinding the cups and cones of bicycle ball bearings, Fig. 199. The traverse of the emery wheel on this machine was controlled by templates so that the profile of the ball race produced could be altered to suit the different designs of different manufacturers. For the internal grinding of the cups extremely small grinding wheels were necessary, rotating at 28,500 rev./min. to achieve the peripheral speed required. The grinding spindle was driven by a lightweight cotton belt.

The production of the pocket micrometer, capable of measuring to an accuracy of one thousandth of an inch, by Brown and Sharpe in 1868 led the way to new conceptions of accuracy and precision of workmanship in toolrooms all over the world. J. W. Roe[10] has described how the features of a screw micrometer for gauging the thickness of plate devised at Bridgeport, U.S.A. in 1867 were combined with the elegant scribing of the graduations on the micrometer patented in France by J. L. Palmer in 1848, thus producing the Brown

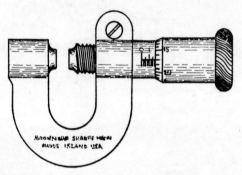

Fig. 200. Brown and Sharpe micrometer.

and Sharpe micrometer, Fig. 200. The idea was not original—William Gascoigne had made a micrometer in England in 1638 (Fig. 93)— and it is known that James Watt had a pocket one in 1772 and Henry Maudslay had used a more accurate bench micrometer in 1800. The value of Brown and Sharpe's achievement in 1868 was that these instruments were produced both accurately and in quantity, that they were handy to use and easy to read and were produced at such a reasonable price that they were soon within the reach of every craftsman. Similar claims have been made for the introduction by J. R. Brown of the vernier calliper in 1851, but this was not so far-reaching in its consequences.

The British took the lead in using electric power for portable tools. In 1887 F. J. Rowan described[11] some portable electric drills used for shipbuilding. They were clamped by electro-magnets to the plate being drilled—a method of clamping that found other applications later. Fig. 201, from Rowan's paper, shows one of the machines he described with a fluted twist drill in position. The need for portable tools in shipbuilding was urgent and other papers of the time describe steam and hydraulic riveting apparatus and a pneumatic caulking tool. Compressed air was being used for portable tools—particularly in mining—by 1860 if not before.

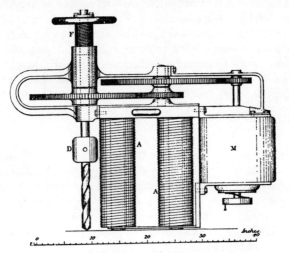

Fig. 201. Rowan's portable electric drill.

The process of feeding a small quantity of sand continuously into a stream of compressed air issuing at high velocity from a nozzle was invented by B. C. Tilghman in the United States in 1871. He called it sand-blasting and it was described to mechanical engineers in England in 1873.[12] It has been used ever since for fettling iron castings and for removing the surface from any hard material, such as cleaning mill scale from steel plate or making ornamental decoration on the surface of plate glass.

Several of the methods now used for welding metals were being tried by adventurous engineers before 1900 but except for forge welding in a coke fire, the modern welding methods were not in general use for manufacturing. Electric arc welding was invented in 1885 when an electric arc was drawn with a carbon rod connected to one side of an electric generator, the other being coupled to the workpiece of iron or steel, when local melting of the metal took place. Metal wire was used in place of the carbon rod about ten years later. The welds cannot have been satisfactory for the process was not properly understood and the conditions for good welding were not being fulfilled. Electric resistance welding was discovered about the same time. Here an arc was drawn between two metal rods connected to an electric generator and when their ends began to melt they were forced together, producing a forge weld. The method was applied to join metal rods together before wire drawing so as to make wire-drawing a fully continuous process. When oxygen became available commercially in the 1880s, it was discovered that steel could be cut with the flame from a hand torch in which the oxygen was mixed with town gas or hydrogen, a method that was made use of by safe-breakers in the last century. For welding steel the use of the blow-pipe was not practical until acetylene became available after 1900.

MACHINES

The place that was occupied by the railway locomotive as the most important mobile machine created in the period of the last chapter was taken by the mechanically-propelled road vehicle in this period. The horseless carriage had been prophesied centuries before; now it became a reality, and at first indeed, it looked like a carriage without a horse, for the engine was small and placed inconspicuously under the body of the carriage.

Steam carriages were made in comparatively small numbers from about 1830 onwards but they were slow, large, and clumsy vehicles like Hancock's (Fig. 182) designed to carry a number of passengers, and though towards the end of the century Serpollet in France and the Stanley brothers in the United States had made light-weight steam carriages, the steam engine with its requirement of a boiler and either a condenser or a constant supply of water, was unsuited to road traction. Thus it was not until the oil engine was available in the 1880s that successful horseless carriages could be made. Before 1900 such vehicles, powered by small petrol motors, had been built by Benz and Daimler in Germany, by Peugeot and Renault in France, by Lanchester and Austin in England, and by Henry Ford in the United States. There were also others in these countries and elsewhere but those mentioned achieved success and popularity for their machines in the next century. Most of them were enthusiasts with some practical knowledge that they had gained as mechanics, but an exception was F. W. Lanchester, a distinguished mechanical engineer before he ventured into automobile engineering and aviation. He was probably the first to apply scientific principles to the design of the self-propelled road vehicle, and his design of

1897 incorporated pneumatic tyres and Ackerman steering, a worm drive to the back axle which had a differential, epicyclic gear box with roller bearings specially made, and a balanced engine placed in the centre of a tubular chassis which was sprung at the front on cantilever springs. Many of these features were to become almost standard practice in automobile design for the next half-century.

The automobile was to undergo further important improvements after 1900, but the road vehicle that achieved finality of design before then was the bicycle. Perhaps the most significant machine of this period, it passed quickly through the various stages of design from the time in 1861 when cranks and pedals were first applied to the wheels of a hobby horse (a two-wheeled vehicle on which a man could sit astride with his feet touching the ground) through the Rover Safety bicycle of 1885 to the Singer Safety bicycle of 1890. The latter (*see* Fig. 245) had curved front forks so that the steering axis passed through the point of contact of the front wheel with the ground; in other respects both machines incorporated all the essential features of the bicycle of today, namely the diamond-shaped tubular steel frame, wire-spoked wheels running in cup and cone ball bearings, two central cranks and pedals driving the back wheel with sprockets and roller chain, a sprung saddle and (after 1888) pneumatic tyres.

The publication in 1875 of Reuleaux's *Kinematics of Machines* was a landmark in the development of this subject, for until then writers of books had treated machines as a whole in considering the motions of their individual parts, and where similar parts recurred in different machines these were called mechanisms. What Reuleaux did was to define a machine as 'a combination of resistant bodies so arranged that by their means the mechanical forces of nature can be compelled to do work accompanied by certain determinant motions'. He then pointed out that the change of position of the parts—that is, their motion—is determined simply by the geometry of the parts; for example, two parts might be a screw and a nut and for these only one motion was possible for each relative to the other. Two such parts he called a pair of elements, and the elements might be combined into a link and the links united into what he called kinematic chains. He showed that by locking one such link a mechanism was obtained and as an instance of the value of this conception to the analysis of the motion of machines, he showed in one chapter that many forms of rotary engine have the same kinematic chain of elements as the common direct-acting crank engine and differ only in the position and manner of the fixing restraint.

Professor Reuleaux elaborated a system of notation for using his analytical methods which assisted in resolving complex mechanisms to a remarkable degree. He also included fluids where they occur in machines; he called them 'pressure organs' and considered their kinematic behaviour to be the reverse of that of tension organs in the mechanism. His book was widely used in all industrialized countries and was translated into many languages, the English translation being by Professor Alexander Kennedy.

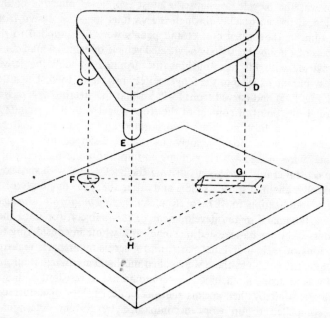

Fig. 202. Kelvin's coupling.

One of the consequences of this interest in kinematics was the birth of the principles of kinematic design of machinery.[13] An example is shown in Fig. 202 of Kelvin's Coupling, a three-legged instrument stand which if once levelled will remain so in spite of changes of temperature that would cause distortion of more conventional supports. The upper platform stands on three legs rounded at their lower ends, C rests in a conical depression F, D in a groove of triangular section G, and E, which is shorter than the other two, on the top of the base plate at H. The upper portion may be removed and when replaced will occupy exactly the same position in space as before, without the need for re-levelling or adjustment. This form of kinematic coupling is widely used on surveying instruments, but the application of the geometrical principles involved was not commonly adopted in engineering machinery until the present century.

There was much pre-occupation with valve gears for steam engines at this time and with graphical methods for representing their motion. Much of this graphical work involved the integration of areas, on the drawing board, a process that was made simpler and quicker by the use of the planimeter (Fig. 203) invented by Amsler, who died

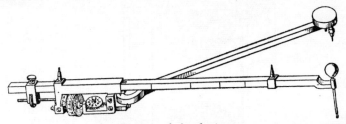

Fig. 203. Amsler's planimeter.

in 1849. With this instrument the area enclosed within an irregular boundary line can be measured directly by moving the pointer by hand round the boundary until the starting point is reached when the area can be read off on the scale of the integrating wheel. It also became fashionable to draw diagrams representing displacements, velocities and accelerations of parts of machinery such as cams, pistons and points on connecting rods and linkages. Likewise, the storage of energy in flywheels was studied by turning-moment diagrams and otherwise, various forms of centrifugal governors were designed for controlling the speed of rotating machinery, and the sensitivity of these devices and their liability to 'hunting' came to be generally appreciated.

The growth of automatic controls was an important aspect of machine development in this period. In 1867 Macfarlane Gray[14] described the automatic steam steering gear (Fig. 204) that he had designed for the giant *Great Eastern* steamship. It proved very successful and gear of this type has been widely used for steam steering ever since; equally important was the automatic hydraulic steering gear[15] devised by A. B. Brown (1870) who established works at Rosebank, Edinburgh, where hydraulic steering gear is still made. Others who invented interesting steering gear for ships at this time were F. E. Sickels of New York (1860), and Joseph Farcot of Paris (1872), who invented the term servo-motor to describe 'a feed-back power-amplifying control system', while Franz Reuleaux, F. Linke and T. Ritterhaus in Germany discussed power relays, and servo-systems for the automatic control of machines in factories.

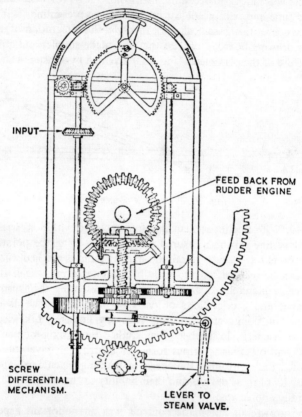

Fig. 204. Macfarlane Gray's steam steering-engine.

The first application of a true servo-mechanism to the regulation of the steam supply to an electric generator was made by P. W. Willans of Rugby in 1882. His apparatus[16] was a solenoid-operated valve system using a two-stage valve not unlike some regulators used on alternators today. Willans discussed the problem of 'hunting' that bothered so many designers of early control devices. Had he known it, the principles involved in overcoming the tendency to instability of these controllers had been fully explained by Clerk Maxwell in 1868 when he had proposed two remedies, the first to introduce friction to oppose the motion of oscillation and the second to introduce an element of anticipation into the action of the controller. The latter method, which is most commonly used today, requires the use of instruments for measuring rate of change that were not available in the last century. Maxwell determined theoretically the amount of sensitivity and anticipation required of a controller to obtain stability in some simple systems, and some years later his work was extended by another Cambridge mathematician E. J. Routh.[17]

Another application of automatic control that began to grow at this time was the control of temperature.[18] It will be recalled (Chapter V) that the first of these was for the temperature compensation of pendulum clocks—Harrison's Gridiron pendulum, 1726. Bimetallic strips were used for this purpose for the next century and in 1830 this principle was used by Dr. Andrew Ure of Glasgow in a device for controlling the temperature of distillation processes. This device he called a thermostat, a word now used to describe all temperature controls. During the next fifty years many other thermostats were patented, particularly in the United States, where ideas for labour-saving were popular. In Britain a new type of thermostat giving very sensitive control at a particular temperature was introduced by C. E. Hearson in 1880.[18] The control element was a metal capsule consisting of two thin sheets of metal sealed together at their edges and enclosing blotting paper saturated with gasoline or a liquid boiling at the temperature desired in the apparatus. When this temperature was reached the liquid in the capsule boiled, causing it to expand and so moving a damper in a chimney to divert some of the hot gases from the lamp (or fire) away from the oven that was being heated. Hearson's apparatus (Fig. 205) was for the artificial incubation of eggs, for which it is still used, but the notion of employing a capsule in this way has found many other applications.

The balancing of locomotives was the subject of controversy in the English technical press during the 1850s. It had been noticed that increasing speeds had accentuated the 'nosing' effects due to unbalanced engine masses. To investigate the matter, oscillation diagrams were taken from working locomotives while they were suspended in the air by ropes and chains, when it was found that the least oscillation occurred when all the rotating masses and two-thirds of the reciprocating masses were balanced, a practice that then became established.

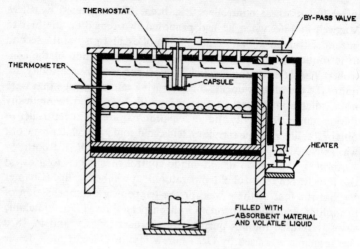

THERMOSTAT

BY-PASS VALVE

THERMOMETER

CAPSULE

HEATER

FILLED WITH
ABSORBENT MATERIAL
AND VOLATILE LIQUID

ENLARGED VIEW OF CAPSULE

INCUBATOR WITH AUTOMATIC TEMPERATURE CONTROL. 1880

Fig. 205. Hearson's incubator.

Following a number of railway accidents efforts were made to improve the braking on railway trains. After extensive trials of different systems in 1875 the Westinghouse Continuous Automatic Compressed Air Brake (Fig. 206) was generally adopted. This braking system utilized a reservoir slung under each carriage in which air at 60 to 100 lb./sq. in. was maintained by a steam-driven compressor. The air was used behind pistons which forced cast-iron brake shoes against all the carriage wheels. The system was quickly adopted both for main-line express trains and for suburban trains where rapid acceleration and deceleration were needed. The conditions to be satisfied were: (1) that the train must be braked by the application of brake shoes to all the wheels of the train, (2) that the brakes must be under the control of the driver, (3) that in an emergency they should be able to be applied by the guard from the rear of the train, (4) that the compressed air supply pipe between the carriages must be flexible, and (5) that the brakes must act automatically on all the carriage wheels in the event of accidental separation of the train. It will be seen from the figure that so long as the pressure of the compressed air in the continuous pipeline was maintained, the diaphragm on each carriage remained horizontal and the brakes off.

If the air pressure in the line was lowered, the non-return valves in each air receiver and above the diaphragm closed and the consequent difference of pressure across the diaphragm caused it to deflect and so turn the valve, thus applying compressed air to the brake cylinders. Restoring the pressure of the compressed air in the pipe line put the brakes off. A similar principle was used in the vacuum automatic brake adopted by many railways.[19]

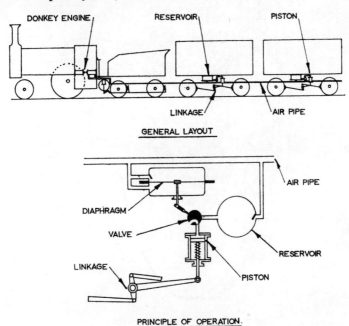

Fig. 206. Westinghouse automatic compressed-air brake.

The development of road vehicles of all kinds led to a number of important inventions of machine elements and components: for example, the steam carriage—and later the automobile—needed for successful operation improved devices that were soon supplied by such items as band brakes, friction clutches of the disc type (Fig. 207), gear boxes with sliding gear changes, and the bicycle required for its evolution the cup and cone ball bearing, the bushed roller chain produced by Renold in 1879 and the pneumatic rubber tyre in which success was first achieved by J. B. Dunlop, a veterinary surgeon of Belfast, in 1888.

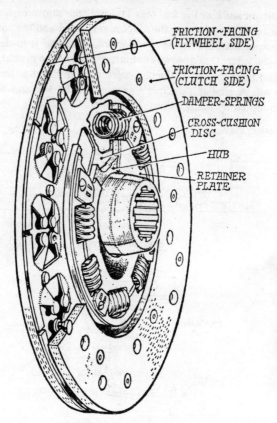

FRICTION~FACING
(FLYWHEEL SIDE)

FRICTION~FACING
(CLUTCH SIDE)

DAMPER~SPRINGS

CROSS~CUSHION
DISC

HUB

RETAINER
PLATE

Fig. 207. Disc type clutch.

The method most commonly used for transmitting power in factory buildings in the 1860s is given by Fairbairn.[20] Wrought-iron overhead shafting was used on which pulleys were mounted where required. The shafting was driven from the engine either by machine-cut gear wheels or by multiple ropes. The final drive to the machinery was usually by flat leather belts with spliced joints. The shafting in some factories might extend in single lines as long as 350 ft. and total two miles in length.

The overhead travelling crane was in use in Whitworth's workshops in London in 1858. A steam engine on the shop floor supplied power through pulleys and belts to a long shaft mounted on bearings on the crane rails and from this shaft power was available for long travel, traversing and hoisting. Soon afterwards such cranes were provided with a small steam engine and coke-fired boiler mounted on the crane gantry. Some of these cranes were still in use in England as late as 1930.

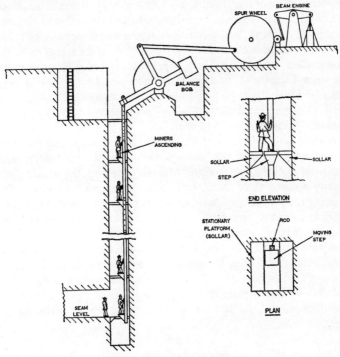

LAYOUT OF TYPICAL CORNISH MAN-ENGINE.(RECONSTRUCTED).

Fig. 208. Cornish man-engine.

A remarkable handling machine that made its appearance in the deep mines of Cornwall was the man engine,[21] Fig. 208, a device for raising and lowering miners—'a pump to pump men.' It originated in the Hartz Mountains in Germany in the 1830s, and bears a resemblance to the *stangenkunst* of Chapter V but is vertical instead of horizontal. As developed on the Continent man engines had double rods and were so greatly elaborated that they became unreliable and highly dangerous, but the winning design of a competition in England in the 1840s, that became known as the Cornish man engine was a simple type, as in the illustration. A single oscillating rod (built up of wood and wrought-iron straps) was raised and lowered through a 12-ft. stroke by a crank (6-ft. radius) and bob at the surface worked by a water wheel or steam engine. Projecting steps were fixed to the rod and spaced equally along its length, and they came to rest opposite platforms or sollars fixed to the wall of the mine shaft. When riding upwards the miner moved from the sollar to the step as the rod dwelt at the bottom of its stroke and stepped sideways to the next sollar above when the rod dwelt again at the top of its stroke (4–6 strokes per minute). There were balance bobs at intervals down the mine to make it steadier. The engine could work in a sloping shaft by fitting angle bobs where the slope changed. The weight of such an engine for a mine of about 300 fathoms was 14 tons unloaded, 24 tons carrying 150 men.

A great saving in time was made with the man engine, of the order of one hour per man per journey, as compared with ladders, and accidents occurred less often. Mortality among the miners was greatly reduced since there were fewer heart and lung cases. The running costs were only one-quarter. However, by 1880, the public attitude to safety was more alert and the man engine came to be considered rather dangerous, and rigid rules for its safety in operation were drawn up. From this time onwards the use of the man engine declined and after an accident in 1919 at the Levant Mine in Cornwall, when the cap at the head of the rod broke, and the falling rod carried away sollars and men with a loss of 31 lives, their use was discontinued, the last one being closed down at Laxey, Isle of Man, in 1920.

The first successful typewriter[22] was developed in Milwaukee, Wisconsin, by C. L. Sholes and C. Glidden in 1868 and built to their order by E. Remington and Sons in the United States. Many others before them had patented typewriting machines of different kinds, without achieving practical success. There was little in Sholes and Glidden's design that had not been anticipated before. What they did was to continue experimenting with a key system and linkages to the type bars, until they had a universal arrangement of keys that enabled operators to type rapidly without the type bars fouling one another. The Sholes and Glidden typewriter became known as the Remington No. 1 in 1876.

FLUID MACHINES

Some remarkable advances were made in the development of fluid machinery during the latter half of the last century. The outstanding contributor was Osborne Reynolds, whose discovery in 1883 of the two modes of motion of flowing fluids made possible the analysis of

Fig. 209. Reynolds's apparatus.

fluid motion with a precision that was impossible before. The two modes of motion are called streamline and turbulent, the former when particles of the fluid follow defined and orderly paths and the latter when the motion of individual particles is quite random and unpredictable. Reynolds illustrated these two modes of motion in the case of water in an elegant experiment (Fig. 209) using a glass-sided tank containing a horizontal glass tube with a bell-mouthed inlet, the tube being fed from the still water in the tank, while a fine

L

filament of ink was introduced into the mouth of the tube. As the water flowed out of the tank through the tube the filament appeared as a streak parallel to the tube axis (Fig. 210*a*) but as the velocity was increased there came a point when the motion suddenly became turbulent so that the filament of ink was diffused into the whole body of

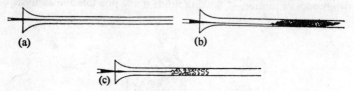

Fig. 210. *Reynolds's number.*

the water in the tube. (Fig. 210*b*). This phenomenon occurred at the 'critical velocity' which is that when the motion changes from the one mode to the other. Reynolds showed that the transfer from the turbulent mode to the streamlined one occurred when the dimensionless parameter

$$\frac{vd\rho}{\mu}$$

had the value 2,000, with a widely varying diameter of tube. (*d* is the pipe diameter and *v*, ρ and μ are respectively the velocity, density and viscosity of the fluid—all in a consistent system of units.) The quantity

$$\frac{vd\rho}{\mu}$$

is now known as 'Reynolds's number' and it is one of the most important quantities in the study of fluid motion.

One of the immediate fruits of this discovery was that it explained the apparently discordant results of experiments in resistance to fluid flow in tubes by Poiseuille (1846) and Darcy (1857), Reynolds was able to show that Poiseuille's experiments had all been carried out with velocities below the critical when the resistance is proportional to v/d^2 which his results confirmed, and that Darcy's experiments had all been made above the critical velocity when he found the resistance to be proportional to v^2/d. The use of Reynolds's criterion in fluid motion has very wide application, not only for fluids flowing in pipes but in aeronautics, and marine work, in heat transfer and combustion, and indeed in any application where fluid flow is concerned.

In 1886 Reynolds presented to the Royal Society his classical paper on the theory of lubrication. This contained an analysis of the experimental results on lubrication obtained by Beauchamp Tower for the Institution of Mechanical Engineers. What Reynolds did was to apply the equations of hydrodynamics to the motion of the oil in the bearing. The analysis indicated that for the maintenance of a continuous oil film the radius of the bearing must be greater than that of the shaft so that the latter could rotate eccentrically within the former by an amount depending on the load, the speed and the viscosity of the lubricant. This work explained for the first time how the oil film in a bearing is maintained and the same method was later applied by Michell and Kingsbury to design suitable forms of thrust bearing.

It was Beauchamp Tower[23] who showed the value of flooding a bearing with oil to reduce mechanical friction and it was he who observed that the journal itself acts as a pump and raises the oil pressure (Fig. 211) so that the oil film is able to support the shaft without its ever touching the bearing.

Soon after these ideas had been made public the system of forced lubrication for high-speed engines was invented by A. C. Pain and incorporated in the enclosed high-speed steam engine introduced by Bellis and Morcom in 1890. In the forced lubrication system an oil pump driven by the engine is provided to circulate the lubricating oil under pressure continuously through the bearings. The system has been applied universally to high-speed machinery ever since. Although Pain was the first to apply this method of lubrication to a reciprocating engine the idea of forced lubrication was anticipated by C. A. Parsons in 1884 when he incorporated a direct-driven screw pump in his first steam turbine to circulate the lubricating oil to the main bearings of the machine.

Osborne Reynolds also made important improvements to the design of centrifugal pumps. In 1875 he patented the multi-stage centrifugal pump, in which a number of pumps mounted together on a single shaft were connected together in series so that a high delivery head could be achieved. He also surrounded the impellers with guide vanes and divergent passages so that some of the kinetic energy of the fluid could be converted into pressure head within the pump, thus improving its efficiency. In 1887 one of his pumps was found to have an efficiency of 55 per cent when running at 1,500 rev./min. and delivering 216 gallons per minute against a head of 150 ft.

Reynolds was also among the first to use a centrifugal pump as a hydraulic brake for measuring power output from an engine by mounting the casing so that the torque or twist exerted upon it could be measured by balancing it against the moment provided by a weight on a steelyard. This idea was first used by Hirn in 1865 but in 1877 an improved type of hydraulic brake was produced by

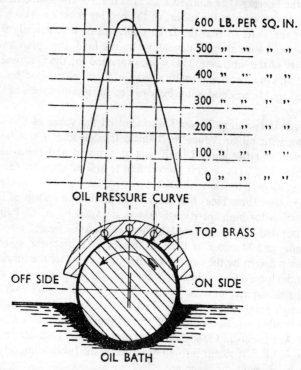

600 LB. PER SQ. IN.
500 ,, ,, ,, ,,
400 ,, ,, ,, ,,
300 ,, ,, ,, ,,
200 ,, ,, ,, ,,
100 ,, ,, ,, ,,
0 ,, ,, ,, ,,

OIL PRESSURE CURVE

TOP BRASS

OFF SIDE   ON SIDE

OIL BATH

Fig. 211. Experiment of Beauchamp Tower.

William Froude, whose design produced a very large resistance for its size. He achieved this by causing the water to circulate within the casing in chambers formed by the vanes of the rotor and vanes on the stationary casing. The forced vortices produced within tended to lower the pressure at the centre, causing instability, until Reynolds suggested drilling small holes through the casing to the centre so that the pressure there should always be atmospheric. The Froude hydraulic brake (Fig. 212), or dynamometer, is still a standard piece of equipment, used for power measurement, particularly for the measurement of very large powers. A single brake may be used to measure as much as 20,000 horse-power.

William Froude, and later his son Robert, achieved fame for their pioneer work in measuring experimentally the frictional resistance of planks and model ships as they were towed through water in a

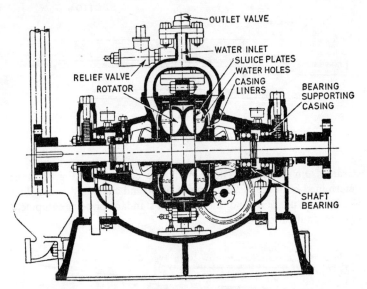

Fig. 212. Heenan and Froude's hydraulic dynamometer.

long tank (Fig. 213) specially designed for the purpose. In 1872, with funds obtained from the Admiralty, Froude built a towing tank 250 ft. long at Torquay, where he and later his son studied experimentally the effect on skin friction and wave formation of the speed, the shape and the finish of the surface being drawn through the water. These pioneer model tests were the basis of what has now become an important part of the science of naval architecture—the estimation before a ship is built of the power required to propel it at different speeds through the ocean.

Soon after Froude began measuring the skin friction of planks in a water tank, another Englishman, H. F. Phillips, began to measure the drag of vanes in a wind tunnel. He described tests[24] made in an 'open' type of wind tunnel with a test section 17 in. square by 6 ft. long, the air flow being induced by a steam injector down-

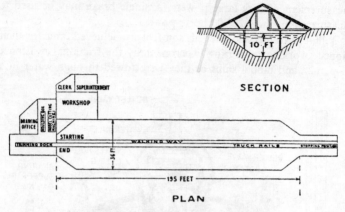

Fig. 213. Froude's towing tank.

stream from the test section. With this equipment (Fig. 214) he determined both the lift and drag of plane and cambered vanes. This was the first use of a wind tunnel, now a standard piece of equipment for aeronautical engineers.

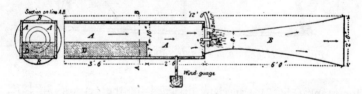

Fig. 214. Phillips's wind-tunnel.

By 1889 the German engineer Otto Lilienthal had taught himself to fly a machine which he called a hanging glider (Fig. 215) and before he was killed in a gliding accident ten years later he had made more than a thousand successful flights with different kinds of experimental gliders. From these flights a great deal of valuable experience was obtained, later to be used when powered flight in aeroplanes began in the next century. An important step in the development of the aeroplane was the invention of the box-kite by L. Hargrave in Australia in 1893. This formed the basis of the design of bi-plane gliders and later of the trussed bi-plane structure used for a number of early successful aeroplanes.

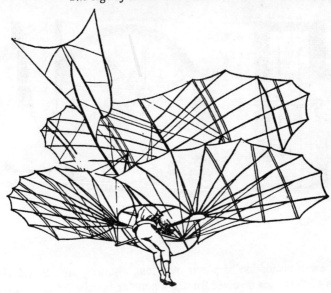

Fig. 215. Lilienthal in his glider.

During the 1860s a change took place in the methods used for ventilating coal mines. Before then the usual method was a furnace at the base of the upcast shaft with the galleries so arranged that air was drawn through them by the draught caused by the furnace. By this time the dangers associated with fires underground were appreciated and alternative means of ventilation came to be adopted. Various types of centrifugal fan were used, of which the Guibal fan (Fig. 216) is typical.[25] The one shown was 17 ft. in diameter and was driven by a steam engine at the top of the upcast shaft.

Another type of mine ventilator was the positive displacement or Roots' blower (Fig. 217)[26] which found many other applications where a somewhat higher pressure was required than could be supplied by a centrifugal fan. The Roots' blower was invented in the United States by two brothers of that name in 1866. For such purposes as blowing a cupola for making iron castings, Roots' blower and the compound centrifugal fan became popular. In the compound fan, two fans mounted on the same shaft were connected together in series in the same way as in the multi-stage centrifugal pump.

The use of compressed air was becoming more general, particularly for tunnelling, mining and for portable tools. It was even used

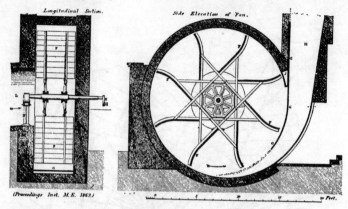

Fig. 216. Guibal's ventilating fan.

for the transmission of power over long distances and the calculation of the losses involved formed a popular exercise for engineering students at the end of the century. In 1878 Peter Brotherhood in England patented a reciprocating three-stage dry air compressor with water jackets, intercoolers and aftercooler originally intended for torpedoes but found to have many applications for industrial purposes. The designed delivery pressures went up to 1,700 lb./sq. in.

The outstanding achievement in water power machinery was the invention of the Pelton wheel in the goldfields of California. Water

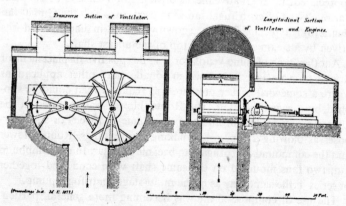

Fig. 217. Roots's blower.

wheels were used for power production, originally with flat vanes and later with hemispherical cups, the water jet impinging on the centre of each cup. It is said that Pelton, who was in charge of one of these wheels during the 1880s noticed that when a cup came loose so that the jet of water impinged on the edge of the cup, the wheel speeded up. Thus originated the water wheel with hemispherical buckets having a central partition, now known as the Pelton wheel (Fig. 218). Very high efficiencies (of the order of 85 per cent or more) have been obtained with Pelton wheels, which are particularly suitable when the water supply is available from a high head.

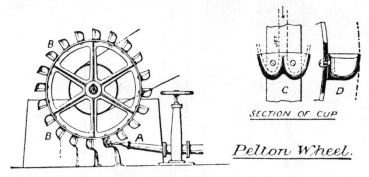

Fig. 218. Pelton wheel.

Reciprocating hydraulic pumps were still in use throughout the period, driven by steam engines, sometimes through a beam. They were particularly favoured for water supply and during this time some of the pumps built for this purpose were very large; for example, for London's water supply in 1855 a steam-driven pump of 380 h.p. was installed having a cylinder 100 in. in diameter by 11 ft. stroke, and some time later another of 880 h.p.

One of the pioneers of hydraulic power at this time was Sir William (afterwards Lord) Armstrong, who devised a hydraulic accumulator which proved to be a most useful adjunct to any installation where hydraulic power was in use on a large scale. In its simplest form the hydraulic accumulator (Fig. 219) consisted of a vertical cylinder containing a weighted piston, the weight and size of the piston being proportioned to equal the hydraulic pressure in the main. This device made it possible to operate cranes, lifts, jacks and other hydraulic devices from the supply, even when the supply pump was at rest, and ensured a constant supply pressure for the machines when the pumps were working. Armstrong was a pioneer in the development of hydraulic systems for the transmission of power which were established in a number of large cities such as London, Glasgow, Hull and Manchester, where the water pressures were usually of the order of 1,000 lb./sq. in. Higher pressures than the supply pressure could

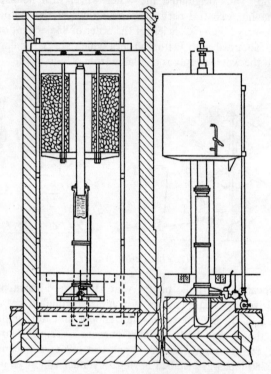

Fig. 219. Hydraulic accumulator.

always be obtained by the use of a hydraulic-pressure intensifier (Fig. 220). By the end of the century exercises on the economics of power transmission over long distances by hydraulic power were part of the training of many young mechanical engineers.

One of the most useful mechanical instruments of this period that has retained its popularity for more than a century is the Bourdon pressure gauge (Fig. 221).[27] Its inventor, Monsieur Bourdon, noticed quite by accident in a workshop in France about 1850 that a circle of tubing which had been flattened, straightened out when an internal

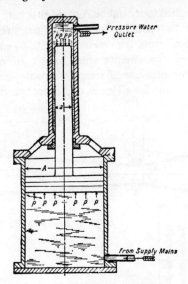

*Fig. 220. Hydraulic pressure
intensifier.*

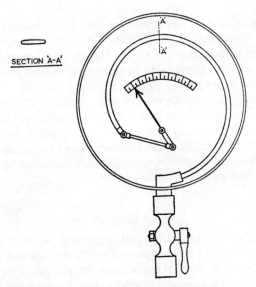

Fig. 221. Bourdon pressure-gauge.

331

pressure was applied to it, and he realized that here was the principle of an instrument that could be used for the measurement of pressure. On many occasions since then men must have owed their lives to the reliability of these instruments which have been made in millions for measuring the pressure of steam, air, water and indeed almost every kind of fluid and for all kinds of different purposes.

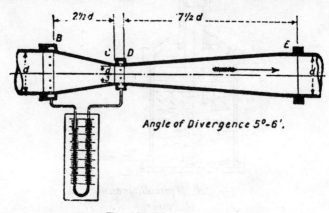

Fig. 222. Venturi meter.

A useful instrument was devised in 1887 for measuring the flow of liquids. Known as the venturi meter (Fig. 222) it was invented by the American engineer Mr. Clemens Herschel, who named it after the Italian scientist Venturi who, a century before, had made experiments on the flow of water through conical pipes. The rate of flow can be determined from the reading of the pressure difference shown on the manometer since the dimensions of the meter are known. It is a simple and satisfactory instrument which depends for its functioning on the truth of Bernoulli's theorem. When a continuous record of flow is required, suitable mechanical attachments can be added to record the reading of the manometer

HEAT ENGINES

The development of the internal combustion engine during this period was one of the most important events that has ever occurred in the history of mechanical engineering. The gun can be regarded as a form of internal combustion engine and for centuries abortive attempts had been made to build a machine to use the energy resulting from an explosion in a cylinder. Huygens and Papin were both among the unsuccessful experimenters in this field and the Abbé Hautefeuille also. A long series of patents for unsuccessful gas-exploding engines were taken out during the next century before success was finally achieved by the Frenchman Lenoir in 1860.* Some measure of his success can be deduced from the fact that three or four hundred of his engines were made in France in the next five years. In appearance (Fig. 223) Lenoir's gas engine resembled a

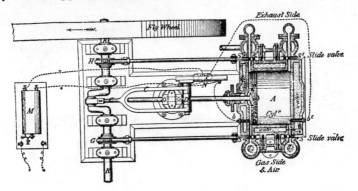

Fig. 223. Lenoir's gas-engine.

double-acting horizontal steam engine. Gas and air were admitted at both ends of the cylinder for a part of the stroke, then fired by an electric spark at atmospheric pressure, and expanded during the remainder of the stroke. There were thus two explosions—one on each side of the piston—per revolution of the crank and after expansion the return stroke was used for exhausting one side of the piston while induction, explosion and expansion took place on the other. The cylinder was water-cooled, and slide valves were used for admission and exhaust. All the Lenoir engines were small, varying from ½ to 3 h.p. and the gas consumption was large. Nevertheless Lenoir produced the first internal combusion engine that was made in numbers for public use, and deserves great credit for having overcome the immense practical difficulties that stood in the way of making a workable machine.

The principal reasons for the uneconomical working of the Lenoir engine and its immediate successors were lack of compression of the

* Among these patentees were John Barber, 1791, Robert Street, 1794, Philippe le Bon, 1799, Samuel Brown, 1823, William Barnet, 1838, Dr. Drake, U.S.A., 1843, Barsanti and Matteucci, 1854.

charge, incomplete expansion and excessive heat loss through the cylinder walls. These faults were realized by another Frenchman, Beau de Rochas, who in 1862 set out the conditions that ought to be fulfilled in an ideal engine and the sequence of operations now known as the four-stroke cycle, more often called the Otto cycle because it was the German, Nicolas Otto, who applied these precepts in the famous Otto silent gas engine in 1876 (Fig. 224). It was a single-acting four-stroke engine, its method of operation being in all

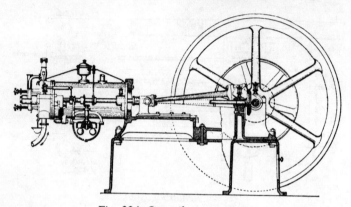

Fig. 224. Otto silent gas-engine.

respects similar to the modern four-stroke engine except that the ignition was by means of a flame carried in a special slide valve. The Otto engine had an immediate success; more than 30,000 engines were sold in the first ten years, and by 1885, 45,000 engines with a total of more than 400,000 h.p. Engines were built under licence in England, France, the United States and elsewhere. Mechanical improvements of various kinds were made, so that the thermal efficiency was more than doubled between 1884 and 1908, typical figures being as follows:

Year	1884	1888	1900	1908
Thermal Efficiency (per cent)	14·3	18·9	25·7	32·2

An important alternative to the Otto four-stroke cycle was invented by an Englishman, Dugald Clerk, in 1880. In the Clerk two-stroke cycle, exhausting of the cylinder takes place at the end of the expansion stroke, and admission of the new charge at the beginning of the compression stroke. Accordingly, a pump or displacer is required to deliver the charge at low pressure into the main cylinder. In the original engine, exhaust took place through ports in the wall of the main cylinder, uncovered by the piston at the end of its stroke. In 1891, by raising the pressure of the pump, Clerk went on to introduce the idea of supercharging. Neither the two-stroke nor the four-stroke cycle has yet established its superiority over the other—both types are used in very large numbers.

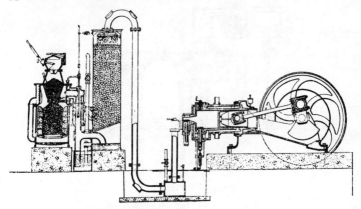

Fig. 225. Gas-engine and producer.

The fuel used in the early gas engines was town's gas but other cheaper forms of gas began to be used in the 1880s—gas from coke-ovens, blast furnace gas and producer gas. Before the end of the period, special forms of gas producer were being designed to be directly coupled to gas engines, the main engine piston being used to provide the suction to draw air into the producer (*see* Fig. 225). These suction gas producers, using inferior grades of solid fuel, were being advertised in 1906 as producing power for one-tenth of a penny per brake horse-power hour. The great majority of gas engines were of the horizontal type.

Some very large engines were built for blowing blast furnaces, using as fuel the blast furnace gas drawn from the top of the furnace. By the beginning of the twentieth century they were producing as much as 2,000 h.p. per cylinder and great batteries of engines were installed in steel works for continuous operation night and day, a measure of the reliability that had by then become taken for granted.

Soon after the gas engine had become a practical proposition, attempts were made to use oil as fuel in place of gas. The first engines, without compression, built by Hock in 1873 in Vienna and Brayton in America at about the same time, achieved little success. After Otto's introduction of compression, the same principle was applied

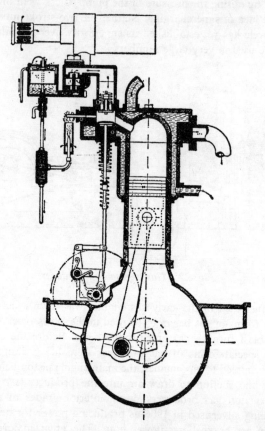

Fig. 226. Daimler petrol-engine, 1897.

to oil engines with improved results, though the difficulties of metering the fuel to the engine, of distributing it evenly throughout the cylinder charge, and of igniting it, remained formidable obstacles for many years. The greatest success was achieved by Daimler, who in 1884 designed a high-speed petrol engine for motor vehicles (Fig. 226) in which the fuel was metered to the engine by a wick carburettor* and ignited by an electric spark. Daimler's engine, which was vertical, was built in sizes from $\frac{1}{2}$ to 25 h.p. with one, two, and four cylinders, and soon ran at 600 rev./min.

In the early stationary oil engines such as the Priestman and Hornsby–Ackroyd, the fuel was sprayed into the cylinder and ignited by means of a hot spot or tube. However, the possibility of igniting

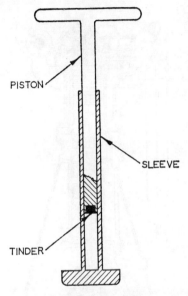

Fig. 227. Fire-piston.

the charge without a hot tube or an electric spark, by the heat of compression in the air alone, was realized by some, and applied in its most successful form by Rudolf Diesel in 1893. This method of ignition had been traditional among Far Eastern primitive peoples for making fire in the form of a device called the fire piston, which had excited interest in Britain as a toy for over a century; in fact a fire piston contained in a walking stick had been patented by Richard Lorentz in England in 1807. The fire piston (Fig. 227) consisted of two parts—a long piston which fitted tightly into a cylinder with a closed end (made of metal, bone, bamboo, etc.)—the former having at its inner end some tinder which became ignited by the sudden heat of compression of the air when the piston was forced quickly down

* For road vehicles the Maybach jet carburettor was soon adopted.

into the cylinder. The piston carrying the lighted tinder at its end was then immediately withdrawn and used to start a fire.

The new feature of Diesel's engine was that the fuel was not introduced into the engine cylinder until the end of the compression stroke, when the temperature of the air had become so high by compression that the fuel ignited spontaneously, having been sprayed in in very

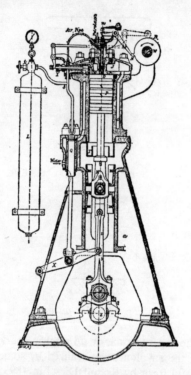

Fig. 228. Diesel-oil motor.

fine droplets through an atomizer. Dr. Diesel also provided his engines (Fig. 228) with air blast injection in which air from another source, at a pressure much higher than the compression pressure, was used to inject and atomize the fuel. He has recorded, in his dissertation 'The Rational Heat Motor' how he aimed to burn the fuel in the cylinder at constant temperature so as to approach the Carnot cycle as nearly as possible. He was never able to achieve this, and combustion in his engine took place approximately at constant pressure; accordingly the cycle of operations in this engine differed in this important respect from the Otto cycle, where combustion takes place while the volume is constant.

To achieve the temperature necessary for spontaneous ignition Diesel was obliged to use higher compression ratios than had been attempted before. Accordingly, pressures were higher and the engine had to be heavier, but as compensation for this it was immediately evident that it had a higher thermal efficiency than any other heat engine. The ideas embodied in the diesel engine were quickly taken up by manufacturers in many countries and its design has continued to improve up to the present day.

The building of stationary gas and oil engines proceeded with astonishing rapidity during this period and no finality was reached as far as design was concerned. Towards the close, there were 42 designs of British gas and oil engines on the market sufficiently well known to be described in a British textbook of the time[28] together with descriptions of 33 German, 28 American, 20 French and 5 Swiss designs.

The introduction of the steam turbine as a practical heat engine by Sir Charles Parsons in 1884 was another of the most important events in the history of mechanical engineering. We have seen in Chapter III how Hero of Alexandria invented his aeolipile or reaction steam wheel about A.D. 100, but this was only a toy. Such also was the impulse turbine devised by Branca, an Italian architect, in 1629. In the last chapter we saw how inventors were struggling with the ideas of both pressure-compounding and velocity-compounding, in the effort to make a successful steam turbine, but though they understood some of the difficulties involved, they did not succeed. Larger types of aeolipile were made by Avery in Syracuse (U.S.A.) from 1837 onwards, by Wilson at Greenock and later by Dr. Gustav de Laval in Sweden. Parsons himself[29] has told us that these machines were used for driving cotton gins and circular saws and he referred to one of Avery's turbines which had a rotor measuring 5 ft. across which reached a speed at the extremities of the rotor of 880 ft./sec. De Laval developed this type of machine to obtain the high speed of rotation needed to drive the well-known cream separator that bears his name. The flyer was mounted on a vertical axis in a casing at the base of the machine, so that it could be directly coupled to the separator spindle. De Laval developed this type of reaction turbine for other purposes, including power production, and the De Laval turbine of 1882, unlike his later impulse turbine, was nothing more than a Hero turbine enclosed in a casing.

Two years later, in 1884, Parsons took out his famous patent and in the same year completed the construction of his first turbine with immediate practical success. This unit developed 7·5 kW. at a speed of 18,000 rev./min. using steam at 80 lb./sq. in. and exhausting to atmosphere. The steam admission was half-way along the rotor, and the steam was divided so that half was exhausted at each end, thus avoiding end thrust. In its passage along the shaft in the annular space between the rotor and the casing the steam streamed through fourteen rings of fixed blades or nozzles, the blades being equally spaced around the circumference, with fourteen rings of moving blades rotating between the fixed blades, forming a continuous ring of jets. The bearings were specially designed to run without producing vibration and they were continuously lubricated with oil supplied by a screw pump driven from the main turbine shaft.

Fig. 229 shows the arrangement of a similar machine, driving an electric generator, described by Parsons in 1888.[30]

The rate at which improvements were made by the original inventor can be seen in the following table comparing the size of the units and their specific steam consumption. By 1900 the steam turbine had nearly outstripped the steam engine in both size and economy.

Year	1884	1892	1900	1913	1923
Power kW.	7·5	100	1,250	25,000	50,000
Steam consumption lb./kWh.	129·0	27·0	18·2	10·4	8·2

The first application of the turbine to marine propulsion was made by Parsons in 1897 in the *Turbinia*, an experimental vessel of 100 tons, fitted with turbines of 2,100 h.p. driving three propeller shafts. It attained the then record speed of 34½ knots.

In a steam turbine the pressure of the steam, instead of being exerted on a piston, is first employed to set the steam itself in motion and the conversion of pressure energy into velocity energy is a necessary preliminary to obtaining effective work from the machine. Parsons achieved his success by dividing up the whole range of expansion into successive steps, so that he prevented the steam from acquiring too high a velocity at any stage.

In 1889 Dr. de Laval introduced a form of steam turbine in which the steam was expanded through a single nozzle, from a region of high pressure to that of low pressure, and in which the speed of the

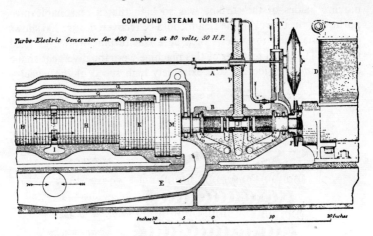

Fig. 229. Early Parsons steam-turbine.

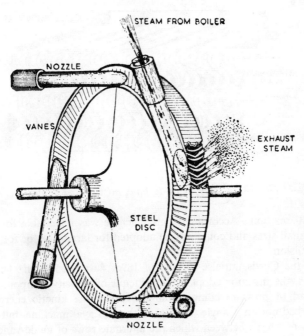

Fig. 230. De Laval turbine wheel.

moving blades of the turbine was so high that a reasonable proportion of the energy of the steam could be recovered. Fig. 230 shows
diagrammatically the wheels of a de Laval turbine with four fixed
nozzles. The novel features of this machine were the divergent shape
of the nozzles inside and the extremely high speed of the wheel—

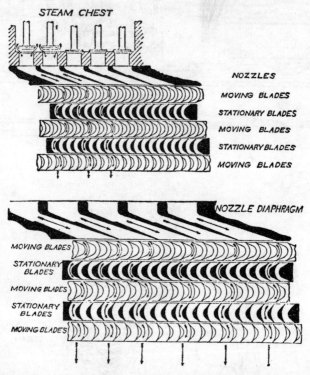

Fig. 231. Curtis steam-turbine, nozzles and blades.

30,000 rev./min. Accordingly this machine was unsuitable for any
but small sizes and could not be adopted for large sizes or for ship
propulsion.

In the Curtis impulse turbine of 1896, developed in the United
States, the inventor adopted the principle of velocity compounding
instead of pressure compounding, whereby the kinetic energy is
absorbed not on a single wheel as in the de Laval machine, but in a
series of wheels between which are alternate rows of guide-vanes to
change the direction of the steam (Fig. 231). These machines were

frequently made with a vertical axis, *see* Fig. 232. Whereas in the early Parsons turbines the effect was produced mainly by reaction, the Curtis turbine is predominantly an impulse turbine. These two types

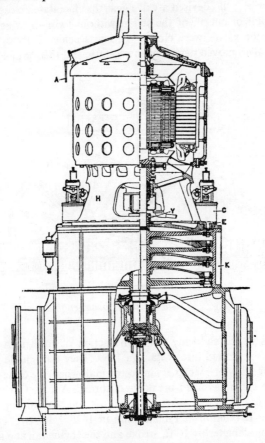

Fig. 232. Curtis steam-turbine arrangement.

of steam turbine have continued to compete with one another where large units for power generation have been required.

This period saw the steam engine rise to its zenith, but before 1900 its position as the most important prime mover was already being challenged by the steam turbine and the internal combustion engine. Nevertheless, reciprocating steam engines, of a bewildering variety of types and sizes were still being designed and built in Britain, the United States, Germany, Switzerland and elsewhere.

An important improvement was patented in 1849 by the American G. H. Corliss, who devised a valve gear that facilitated close control of the point of cut-off of the steam without 'wire-drawing' of the steam as the valves were closed. This was achieved by cylindrical rocking valves moved partly by an eccentric and then tripped so that

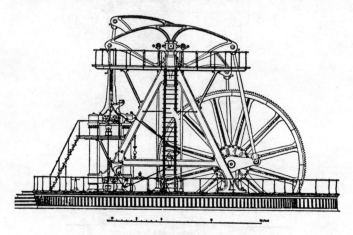

Fig. 233. Giant Corliss beam-engine.

the valves closed suddenly under the action of springs. A Corliss engine was imported into Britain in 1859. They were made in large numbers for driving machinery in cotton mills where the close governing of the engine avoided breakage of thread. A large Corliss beam engine (Fig. 233) was used to drive all the machinery at the Centennial Exhibition at Philadelphia in 1876.[31] Its two cylinders were 40-in. diameter by 10-ft. stroke and the engine ran at a speed of 36 rev./min. The flywheel was 30 ft. in diameter.

The Corliss valve gear was also applied to direct-acting vertical and horizontal engines built in a number of countries because it was both economical and easily controlled.

Many different types of high-speed steam engines were devised. They were in demand for driving electric generators for lighting, for auxiliary machinery on ships and for driving machines in factories. One of the most successful of these was the central valve engine patented by P. W. Willans in England in 1884. In this machine

(Fig. 234) the cylinders, which were single-acting, were arranged co-axially one above the other with the high pressure at the top, and the piston valves were all mounted on the same rod, which was concentric with the engine pistons and was worked by an eccentric in the

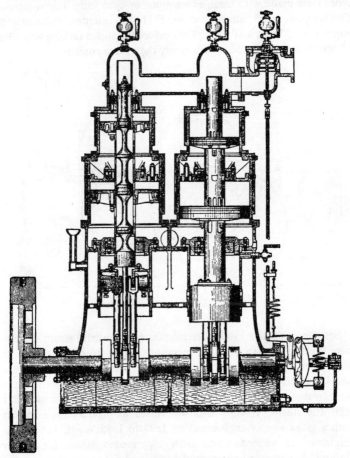

Fig. 234. Willans central-valve engine.

centre of the crank pin. As the cylinders were single-acting there were no reversals of load on the bearings and so the engine was very quiet in operation. It was splash lubricated and the centrifugal governor was mounted on the end of the crankshaft. The engine shown is a compound engine, which means that the steam was expanded twice, first in the upper two high-pressure cylinders from the boiler pressure down to an intermediate pressure and then in the lower two cylinders down to the pressure in the condenser. In the pursuit of better economy some of these engines and those of other types were made with three expansions, or even four. Triple-expansion engines were introduced in 1871 and quadruple-expansion engines in 1875. It was a long time before the steam turbine was able to produce power more economically than these engines.

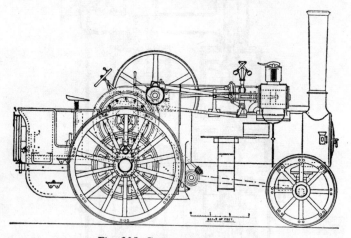

Fig. 235. Steam traction-engine.

Between 1850 and 1900 more efficient and powerful steam engines were built for use in ships, in locomotives and on road vehicles as well as for stationary purposes. In 1881 the steamship *City of Rome* was launched with engines totalling 12,000 i.h.p. and as the years went by engines of more power were built to enable ships to go faster still. The steam pressure used was raised to produce more power with a given size of engine and by 1893 to 1895 W. H. Patchell in England was experimenting with superheated steam for power stations. Superheated steam had been used for some years on the small steam engines of motor cars and trams, particularly in France where the Serpollet steam car with its flash boiler was a popular road vehicle at the end of the century.

One of the steam vehicles to last the longest on the roads was the steam traction engine used for heavy haulage, on fair grounds, and for ploughing and many agricultural purposes. Fig. 235 shows a typical vehicle of this type built by Dodman in England in 1888. The single-cylinder engine of 8½-in. bore and 12-in. stroke developed 30 h.p. when the crankshaft was making 155 rev./min. and as shown on the drawing the drive was by gearing through a countershaft carrying a differential and then to the back wheels. There were two speeds providing road speeds of 2 miles/hr. and 4 miles/hr. respectively. The development of the English traction engine has recently been reviewed by R. H. Clark[32] in a book of that title containing nearly six hundred illustrations.

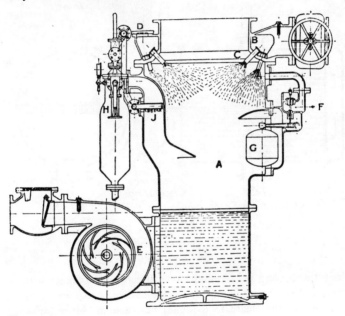

Fig. 236. Jet condenser.

Traction engines, like locomotives, discharge their steam to atmosphere up the chimney, but where economy is an important consideration, as for marine engines and large land engines, a condenser is fitted so as to obtain the lowest possible exhaust pressure by condensing the steam as it leaves the cylinder. It will be recalled (Chapter VI) that James Watt condensed the steam from his engines in a separate condenser where it came into direct contact with cold water. Condensers of this type—jet condensers (Fig. 236)—continued to be used throughout this period. Exhaust steam from the engine or turbine entered at the top where it was met by a fine spray of cold water. The condensed steam or condensate was necessarily mixed with the cooling water and both were pumped away together by the centrifugal pump marked E. With this arrangement the condensate was lost, also all the air in the cooling water had to be removed by the steam ejector H, or by an air pump. Both these drawbacks were overcome by using a surface condenser (Fig. 237)

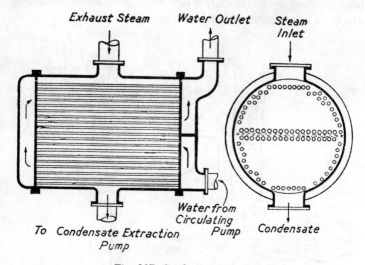

Fig. 237. *Surface condenser.*

where the cooling water never came directly into contact with the steam and so the air pump or ejector had only to remove the air contained in the steam to maintain the vacuum. More important still, the condensate was not lost but could be returned to the boiler, thus reducing the amount of make-up water needed for the boiler feed.

For steam raising, the Lancashire boiler (Fig. 238) was the most popular during this time. Patented by Sir William Fairbairn in England in 1845, it was in all essentials similar to the earlier Cornish boiler (Fig. 176) except that it had two furnaces instead of one, devised, so Fairbairn stated[20] 'so that the furnaces could be fired alternately to prevent the formation of smoke'. Its principal merit compared with the Cornish boiler was that it had a greater heating surface and so could produce more steam.

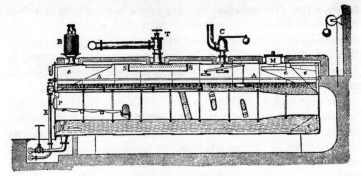

Fig. 238. Lancashire boiler.

The necessity for providing a large heating surface for steam raising had been one of the principal attractions of the water tube boiler for a century before a really successful design was patented in the United States in 1867 by G. H. Babcock and S. Wilcox. An early design of the Babcock and Wilcox boiler which still retains its popularity all over the world is shown in Fig. 239. The water rose up the inclined tubes both by natural convection and because the formation of steam bubbles lessened the density of the contents of the tubes at their upper ends. Steam was separated from the water in the top drum and the water returned to the bottom of the sloping tubes to rise again. This is but one example of many designs of water-tube

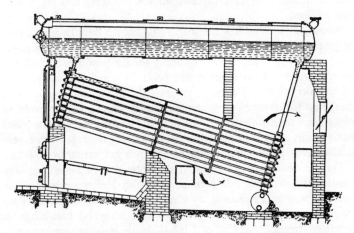

Fig. 239. Babcock and Wilcox water-tube boiler.

boiler, a type of steam raiser that has gained in popularity ever since it was introduced. It is inherently safer than a shell type boiler because if one of the tubes in contact with the fire bursts, it is seldom that anyone is hurt, whereas the explosion of a shell boiler may be calamitous; also, the weight of water in this type of boiler is smaller and so steam can be raised more quickly.

The pioneer work on hot-air engines of Cayley, Stirling and Ericsson during the first half of the nineteenth century was followed during the second half by the manufacture of small engines in very large numbers to designs which incorporated some changes but of which only one was a radical improvement. This was the Rider engine invented in the United States in 1876 and built in three sizes of $\frac{1}{4}$, $\frac{1}{2}$ and 1 h.p. It was also made under licence from 1877 in England, where a thousand engines were sold before the end of the century. The important innovation in this design was that there were two pistons in two separate cylinders, the pistons being connected to cranks which were about 90 degrees out of phase with one another so that the movement of the working fluid back and forth between the hot and cold spaces was caused by the movement of both pistons, neither of which could be called the displacer. [This idea was revived in a further development some seventy years later.] A regenerator was incorporated consisting of a number of cast-iron plates.

In 1872 John Ericsson built a hot-air engine to work from the heat of the sun, but the reflector used was too small and the project was a failure.

About this time an English firm in Manchester built large numbers of small Stirling engines—mainly for pumping water—in three sizes: $\frac{1}{16}$, $\frac{1}{8}$ and $\frac{1}{4}$ h.p. They were very heavy for the power produced, the specific weight of the smallest size being $7\frac{1}{2}$ tons/b.h.p. At the end of the period a still smaller engine—the Robinson engine—was marketed for domestic purposes and many thousands were sold in two sizes, 1 man power and 2 man power. The working piston worked horizontally and the displacer piston, containing the regenerator, vertically (*see* Fig. 240). It could be heated by a gas burner or oil lamp.

On the continent of Europe a domestic size of air engine was made with the two cylinders in line—the Heinrici motor. This machine was tested recently by Finkelstein[33] and found to have an output of 0·028 b.h.p., the specific weight being 1·3 tons/b.h.p. A larger continental machine of 2 b.h.p., made in great numbers, was the Lehmann engine which is shown diagrammatically in Fig. 241.

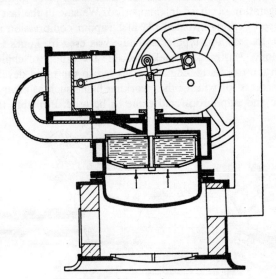

Fig. 240. Robinson's hot-air engine.

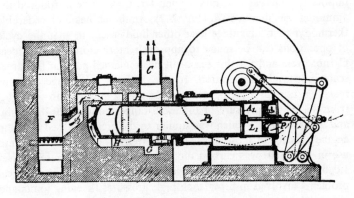

Fig. 241. Lehmann's hot-air engine.

Most of the important events that have taken place in the history of refrigeration occurred in this period. We saw in the last chapter how Jacob Perkins invented the first vapour compression machine in 1834. In 1857 this type of machine was used by James Harrison of Geelong, Australia, to make possible the export of chilled meat through the tropics to England. This led to the rapid development of his process for which other working fluids such as ether, methyl chloride and sulphurous acid came to be used. Probably the most

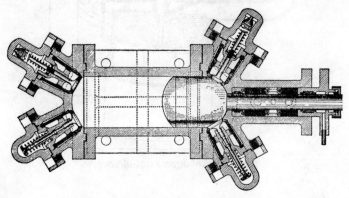

Fig. 242. Linde's ammonia compressor.

important step was that taken when Dr. Carl Linde introduced the ammonia machine in 1876 (Fig. 242). Ammonia has a considerable thermodynamic advantage over other fluids used, in that the cycle of operations can be made to approach more closely to the ideal Carnot cycle and for this reason it has remained popular for large-scale industrial refrigeration. It also had the advantage that for reaching normal freezing temperatures (not lower than $-4°$ F.) from $86°$ F. the pressure of the cycle ranges between 20 lb./sq. in. and 170 lb./sq. in., which was similar to the pressures used in steam engine practice in those days. Another refrigerant which came into use in the 1880s was carbon dioxide. With this a pressure of over 1,000 lb./sq. in. occurs at the high pressure side, but the mechanical problems involved in using such high pressures (keeping joints and glands tight) were successfully overcome and such plants have continued to be favoured for marine work because they are more compact for a given output and any leakage is likely to be less dangerous.

Two other types of refrigerator made their appearance about the same time—the ammonia absorption machine invented by Carré in 1860 and the air refrigerator first invented by Gorrie in the United States in 1845, and improved by Kirk in 1862 with the addition of a regenerator so that it became a reversed Stirling air engine. There were also the various open-cycle air machines first proposed by Lord Kelvin and Professor Rankine about 1852 and made in practical

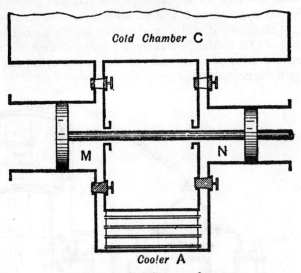

Fig. 243. Bell–Coleman refrigerator.

form by Giffard in 1873 and by Bell and Coleman before 1880. The feature of particular interest about the open-cycle machines of which Bell–Coleman is typical, was that the throttle valve of the Perkins cycle was replaced by an expansion cylinder, so that some mechanical work could be recovered from the working fluid as its pressure fell from the upper pressure to the lower. The arrangement is shown in the diagram, Fig. 243, where M is the compression cylinder and N is the expansion cylinder. Some power had to be provided at the piston rod to make up for the losses. During operation air was drawn from the cold chamber, compressed at M, cooled at A, then expanded while doing work at N so that its temperature fell before it was returned to the cold room. With a compression pressure of 4 atmospheres, expanded down to one atmosphere, the air might return at a temperature as low as −100° F., though the temperature

M   353

of the cold room would never be so low because of heat leakage through its walls.

In the ammonia absorption machine invented by Carré, the ammonia was absorbed in water in the circulating system. It was then boiled off by the application of heat, condensed into liquid anhydrous ammonia in another vessel—the condenser; the liquid was then allowed to evaporate in the next vessel, the evaporator, where heat was absorbed from the surroundings; the ammonia gas was then reabsorbed in the absorber and the solution pumped round again. Fig. 244 shows an external view of such an absorption refrigerator installed in 1876 in Meux's brewery in London where it was in use for more than 20 years. It could produce a ton of ice per hour, though it was normally used for cooling the wort.

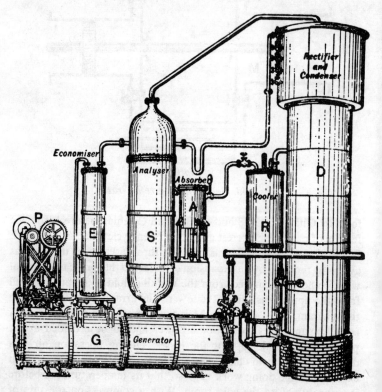

Fig. 244. Carré's ammonia-absorption refrigerator.

REVIEW

As we approach the present day, it becomes progressively more difficult to appraise engineering achievements. Nevertheless, we can select—as the most important innovations in machinery that appeared' in this era—the bicycle, the steam turbine, the internal combustion engine and the automobile. The bicycle, appropriately dubbed 'The Iron Steed', was the first example of a machine that was strong

Fig. 245. Singer safety bicycle.

without being heavy, and it enabled its rider to travel on land by muscular effort two or three times as fast as he could go on foot. It showed a mastery of the problem of mechanical friction between rubbing surfaces, a problem that had become so pressing that the Institution of Mechanical Engineers in England had sponsored research into the lubrication of bearings, the outcome of which had been the historic and brilliant interpretation of Osborne Reynolds. Nevertheless the solution to the problem of mechanical friction found in the bicycle—the cup and cone ball bearing—had no direct connection with Osborne Reynolds's work on lubrication. The bicycle was the creation of the practical man and in the space of a few years a design was produced (Fig. 245) that was so acceptable as to be hardly capable of improvement over the succeeding seventy years during which hundreds of millions of bicycles have been manufactured.

The case of the steam turbine was very different. Even today it is being improved so fast that as soon as each new design has been produced it is already obsolete. The basic principle of the modern steam turbine is still as it was conceived and set to work by Sir Charles Parsons in 1884, but by altering the design, the size, the speed and the steam conditions it is possible to produce vast differences in the output and efficiency of the steam turbine and these changes are continually being made. It is unlikely that the steam turbine could have been invented by a practical man—still less likely to have been improved as it has been. A problem par excellence for the academically-trained engineer, outstanding contributions to its early development were made by men with distinguished academic records—Parsons in England, de Laval in Sweden, Curtis in America, Zoelley in Switzerland and Rateau in France, all engineering scientists. (The considerations involved in designing a steam turbine are largely of a scientific kind. They are not simple, needing to be tackled by people with scientific training and for this reason they have been set as problems for college students and are particularly suitable for written examinations.)

The invention and early development of the internal combustion engine was in its early stages the work of professional engineers such as Dr. Otto, Dr. Diesel and Sir Dugald Clerk but many of the important improvements that made it a cheap and workable power unit have been added by practical men. The joint efforts of both have continued to this day. The application of the internal combustion engine to road transport was almost exclusively the work of practical men particularly in Britain and America where some of those who achieved the greatest success were Ford, Austin and Morris, who started as fitters and had very little formal technical training as engineers. The outstanding exception was F. W. Lanchester whose original contributions to the engineering design of the automobile were greater and more enduring than any of those of his contemporaries.

An innovation of far-reaching importance to the future of mechanical engineering began when T. A. Edison established in 1882 at Pearl St. in New York City the first power station to supply electricity to the public so that power could be purchased from a public supply main like water or gas. Edison was one of the most prolific inventors of his day—the gramophone, the cinematograph and the incandescent filament lamp are three of the best known of his inventions (which hardly come within the scope of this history)—and we must salute him for his vision and his skill in promoting and setting to work the first generating station to supply the public with power and light. Before that time all power and electric light had to be supplied from a prime mover on the premises.

Most of the early power stations were equipped with reciprocating steam engines to drive electric generators. These engines had been greatly improved since the days of the beam engine or the first engines with horizontal or vertical cylinders, and the vertical engine was by now the most popular, having proved itself for reliability in marine propulsion. The high-speed reciprocating engine such as the Willans (Fig. 234) with splash lubrication, and the Belliss with forced lubrication, had been designed to provide the higher speeds required for the direct driving of dynamos, but for the larger units, the slower speed triple-expansion engines with the cylinders mounted vertically at the top of an A-frame were frequently used with a multiple rope drive to the electric generators. This was the most popular form of marine engine, because it was cheap to build, was quiet and reliable, very economical and required no special skilled attendance. (These considerations still applied forty years later when such engines were built in quantities for the Liberty ships by the United States during the Second World War, 1939–45, though by then the building of reciprocating steam engines had practically ceased except for locomotives.)

Many eminent engineers misjudged the future course of events in power production, believing that the prime mover of the future was the gas engine. Their views were based on the much higher thermal efficiency of the gas engine (about 35 per cent) compared with the steam engine (about 23 per cent). Gas was produced from small coal and other waste fuel very cheaply (Tangye of Birmingham advertised power at less than a tenth of a penny per b.h.p./hr. from their suction gas engines in 1906) and gas engines were chosen as the appropriate prime movers for the power station of the city of Belfast though they were scrapped in favour of steam turbines some years later. It was not generally realized at that time that new steels for turbines, boilers and piping, would be developed to enable higher temperatures and pressures to be used in steam cycles, and further that feed-heating, air pre-heating and other new innovations would be developed to improve the efficiency of steam stations. Only a few perspicacious individuals such as Sir Charles Parsons then realized that the steam turbine would have to be made in very large sizes to reduce internal losses to a small fraction and that then its efficiency would exceed that of the reciprocating engine. He achieved this in the next century.

Steam power stations could not have been a success without the improvements that had taken place in the design of steam boilers. This in turn was due in a measure to the establishment of regular inspection of boilers by independent authorities, providing competent engineers who analysed and reported the causes of failures. The frequency of boiler explosions was lessened during the period though many more installations were in use in 1900 than in 1850. Other factors that contributed to greater safety were improved instruments such as pressure gauges and level gauges and the adoption of the water tube boiler for steam raising at higher temperatures and pressures.

Much of the development of mechanical engineering was now concerned with the carrying out of processes such as steam raising, making gas from coal in a gas producer, making electricity or making things cold in a refrigerator. More time was devoted to studying the best way for carrying out the process, whereas previously it was enough to get the process to work at all. Now a type of mechanical engineer who spent his time studying and analysing the operation of a process began to emerge, as the process or production engineer. A leading pioneer in this movement in the United States was F. W. Taylor[34] who towards the end of this period was studying the process of cutting metals, which led him to the discovery of high-speed cutting steels. Taylor also studied other manufacturing processes—even simple ones such as loading pig iron onto trucks by hand and shovelling sand, and found that great savings of labour could be achieved by a scientific study of the process to determine the best way of doing it. Other manufacturing processes that had a considerable effect on the history of mechanical engineering were steel-making—both in the Bessemer converter and the open hearth process, the process of cutting steel with an oxy-gas torch, the sandblasting process for cleaning iron castings and the process for making silicon carbide for synthetic grinding wheels. Most of these processes were basically chemical or metallurgical and their successful development required a knowledge of chemistry or metallurgy which many mechanical engineers did not possess, so that their efforts sometimes resulted in failure. Many did not realize the extent of their own ignorance but those who did sought the collaboration of scientists, often surreptitiously, being ashamed to admit that their knowledge might be insufficient to enable them to do their work properly.

The industrial prosperity of this era in Britain depended less on improved methods than upon manufacturing more goods in the same way and with the same kind of tools and skill that had been used in the previous generation. In machine tools and textile machinery no innovations of importance emerged from England at this time, in marked contrast to the situation in the U.S.A. where new methods of making gear wheels and grinding wheels were being devised and firms like Brown and Sharpe and Pratt and Whitney were building and selling machines that could finish components to fine tolerances without requiring the manual skill from the operator on which the British reputation for fine workmanship depended. The engineers of Brown and Sharpe studied the processes of manufacturing sewing machines, for which they had a contract, to such good purpose that they were persuaded to devise new machine tools on which precision components could be made with comparatively unskilled labour.[2] A minor mechanical marvel of this period was the pin tumbler cylinder lock—an entirely different type from any previous lock, and virtually unpickable—invented by Linus Yale Jnr. in the United States in 1865. Its design was adapted in order that mass-production methods and machinery could be used, for the first time, to produce articles that were *not* identical, in fact each separate lock required a different key. Its security and cheapness, since it did not depend on the handiwork of the skilled locksmith, resulted in its continued popularity for the next century.[35]

In the design of machines there were some curious contrasts. For example there was the light sophisticated design of the safety bicycle (Fig. 245) and the heavy crude conception of the man engine (Fig. 208) surely one of the most dangerous machines ever invented. Another primitive machine was the steam traction engine (Fig. 235) which must have made a strange contrast with the steam turbine (Fig. 229) when Parsons loaned a small steam turbine driving a dynamo mounted on a traction engine to provide electric lighting at a fair in Gateshead some time before 1890. The events in the field of theory of machines that appear to have had most significance for the future were the design of the automobile, the balancing of engines and the extension of automatic control to the steering of ships and the regulation of temperature. Significant also were the theories about kinematic design being developed by Reuleaux and others that were to bear fruit in the next century when applied to the design of measuring instruments, machine tools and mechanisms generally.

Most important advances in fluid machines were made in laying the foundations of the science of fluid mechanics, on which such an elegant superstructure was to be erected in the twentieth century by Prandtl and his collaborators, who based their work on the concepts of fluid motion first enunciated by Reynolds. The discovery of the two modes of motion of a moving fluid—laminar and turbulent— was possibly the most important event in the study of fluids since Archimedes. The application of these theories to the lubrication of bearings was also a 'break-through' from the impasse of empirical data about friction that had been so confusing until the elegant experiments of Beauchamp Tower were so beautifully explained by Osborne Reynolds.

The experimental investigations of drag made by Froude in his towing tank and by Phillips with his wind tunnel provided the basis for the design of gliders and aircraft. All these experiments culminated in the first aeroplane flight by the Wright brothers in the United States in 1903. The use of models for the study of fluid motion had by then been firmly established and this method of experimenting to obtain data for use in designing full-size machinery has, since then, saved engineers from many costly failures.

The water wheel had now followed the windmill into the limbo of obsolete machines, and its place had been taken by the water turbine, an additional type in this period being the Pelton Wheel, one of the examples of a mechanical invention being due to observation of what took place during an accident.

The important science of strength of materials was still being advanced almost entirely by continental engineers in France and Germany. An understanding of this subject, essential if large machines were to be designed to carry heavy loads with safety (as the speed of machinery increased), became still more important, and calculation of the stresses imposed on components was required since when they failed in service it did not always follow that making the parts thicker or heavier would effect a cure.

Young engineers in America were taught the continental methods of calculation by many gifted emigrants who took part in the upsurge of technical education that occurred in the United States. In Britain the teaching of the principles of mechanical engineering was being developed in the universities, where engineering was attracting a small but increasing band of students who were advised to conceal their college education from their employers when they left the university as it might—if it were known—prejudice their chances of employment, since an open contempt for learning existed among many practical engineers.

The textbooks of Professor Rankine on the steam engine and other prime movers and on applied Mechanics were widely used in all English-speaking countries. They were rapidly followed by other authors on hydraulics, mechanism and thermodynamics. Reuleaux's book on the theory of mechanism and Zeuner's book on Thermo-dynamics were both translated into English, and many of the English professors produced their own textbooks such as Perry's *Steam Engine*, Unwin's *Strength of Materials*, Goodman's *Applied Mechanics*, and Dunkerley's *Hydraulics*. Some curious features of the education of mechanical engineers at this time were the comparative neglect of metallurgy as part of the engineer's training (he was taught very little about the properties of the materials he was to use in designing his machines), the complete lack of any teaching of control theory, and the belief in Britain that engineering students should learn about tools and machine tools, not from lectures or reading but by using tools with their own hands, being required to spend many hours and days learning to chip and file a surface flat and to make a forge weld in a blacksmith's fire, occupations that were more proper for an artisan than a professional man. To expect a professional engineer to become a skilled craftsman as well as a scientist and a mathematician must have dissuaded many an enthusiast from taking up mechanical engineering and also probably contributed to the disdain with which mechanical engineering is still regarded in Britain by schoolmasters and members of the other professions.

Before this period professional engineers usually received their training as premium apprentices either indentured to a manufacturing firm or to a consulting engineer, and they learned how to do the work of a mechanical engineer by helping one who was already qualified and doing such work. This method of learning by doing was some-times very successful but often it failed because the apprentices were not given sufficient opportunities of doing responsible work or received no instruction or were exploited in other ways. For the professional part of the engineer's training the apprenticeship system was gradually displaced during this period by formal college education where engineering students were taught machine drawing and design, the testing of materials and machines in the laboratory and exercises in solving simple engineering problems besides basic instruction in Chemistry, Physics, Mathematics and Mechanics. The emphasis on machine drawing as the language in which the mechanical engineer gave his instructions was justified since it was necessary that a generally accepted code of practice should be followed so far as machine drawing was concerned. What seems to have been ignored was that when instruction in design was moved from the works or the consulting engineer's office, the instructors were no longer in touch with the latest ideas and practices of the profession and so the instruction designs of the students were often out of date. The same may be said of the testing of materials and machines which to this day forms the major part of the laboratory work of a student in mechanical engineering. When such instruction was given by a consulting engineer to his pupil he was mainly concerned to teach him how to protect the interests of the client and it seems that mechanical engineers in Britain—such men as Professors Kennedy, Unwin and Ewing who set up the early laboratory courses, must have had in mind that their students were going to become consulting engineers. Now that there are relatively fewer consultants, it is time that the nature and purposes of laboratory instruction should be reviewed.

The results of extending technical education in Britain and the United States did not really emerge until after 1900. In Britain it could perhaps be described as 'too little and too late'. Indeed it is doubtful whether engineering education at the universities could possibly have met the immediate need of industry, which was for successors to the brilliant innovators of the previous period. We record with regret the closing down of the famous firm of Maudslay Sons and Field in 1900 and of James Watt and Company a few years later. The skills that had been acquired and brought together in these establishments for over a century might have been conserved for the further advancement of mechanical engineering, but the founders had taken no steps to provide for a succession of innovators to follow them.

REFERENCES

(for abbreviations see List of Acknowledgments)

1. Bishop, P. W., 'The Beginnings of Cheap Steel', *U.S. Nat. Mus. Bull.*, 218, No. 3, p. 27. Smithsonian Institution, Washington, D.C., 1959.

2. Strassmann, P. W., *Risk and Technological Innovations*. Cornell U.P., 1959, chapter 2.

3. Jobson, R., 'Improved System of Moulding and Casting', *Proc. I.M.E.*, 1858, p. 14.

4. Waterhouse, R. E., *A Hundred Years of Engineering Craftsmanship, 1857–1957*. Tangyes Ltd., Birmingham.

5. Cochrane, J., 'Drilling Machines', *Proc. I.M.E.*, 1860, p. 201.

6. Nasmyth, J., Autobiography (ed. S. Smiles). John Murray, London, 1891.

7. Woodbury, R. S., *History of the Milling Machine*. M.I.T.P., 1960.

8. Woodbury R. S., *History of the Gear Cutting Machine*. M.I.T.P., 1958.

9. Woodbury, R. S., *History of the Grinding Machine*. M.I.T.P., 1959.

10. Roe, J. W., *English and American Toolbuilders*. Yale U.P., 1916.

11. Rowan, F. J., 'Electro-magnetic machine tools', *Proc. I.M.E.*, 1887, p. 323.

12. Newton, W. E., 'Tilghman's Sandblast Process', *Proc. I.M.E.*, 1873, p. 260.

13. Pollard, A. F. C., 'Kinematic Design in Engineering', *Proc. I.M.E.*, vol. 125, 1933, p. 143.

14. Gray, J. Macfarlane, 'Steam steering engine in *Great Eastern* Steamship', *Proc. I.M.E.*, 1867, p. 267.

15. Conway, H. G., 'Origins of Mechanical Servo-mechanisms', *T.N.S.*, vol. xxix, 1953–4 and 1954–5, p. 55.

16. Willans, P. W., 'Electrical Regulation of speed of Steam Engines for Driving Dynamos', *Proc. I.M.E.*, 1885.

17. Macmillan, R. H., *Automation*. C.U.P., 1956.

18. Ramsay, A. R. J., 'The Thermostat or Heat Governor', *T.N.S.*, vol. xxv, 1945–6 and 1946–7, p. 53.

19. *The Vacuum Automatic Brake*. The Locomotive Publishing Co., London, 1921.

20. Fairbairn, W., *Treatise on Mills and Millwork*. London, 1864.

21. Tew, D. H., 'Continental Origins of the Man-engine', *T.N.S.*, vol. xxx, 1955–6 and 1956–7, p. 249.

22. Richards, G. T., *The History and Development of Typewriters*. Sc. Mus. Pubn., H.M. Stationery Office, London, 1948.

23. Tower, Beauchamp, 'Five Reports of the Research Committee on Friction', *Proc. I.M.E.*, 1883–91.

24. Phillips, H. F. (under anon.), Various articles, *Engineering*, 1885.

25. Swindell, J. S. E., 'Guibal's ventilating fan at Homer Hill Colliery', *Proc. I.M.E.*, 1869, p. 78.

26. Carbutt, E. H., 'Roots' mine ventilator', *Proc. I.M.E.*, 1877, p. 92.

27. Cowper, C., 'Bourdon's Pressure Gauges', *Proc. I.M.E.*, 1852, p. 141.
28. Donkin, B., *Gas Oil and Air Engines*. Charles Griffin and Co., London, 1911.
29. Parsons, G. L. (editor), *Scientific Papers of Sir Charles Parsons*. C.U.P., 1933.
30. Parsons, Hon. C. A., 'Compound Steam Turbine and Turbo-electric Generator', *Proc. I.M.E.*, 1888, p. 480.
31. Article in *Engineering*, 19 May 1876
32. Clark, R. H., *The Development of the English Traction Engine*. Goose and Son, Norwich, 1960.
33. Finkelstein, T., 'Air Engines', *The Engineer*, April 1959.
34. Taylor, F. W., *Scientific Management*. Harper, New York, 1911.
35. Giedion, S., *Mechanization Takes Command*. O.U.P., New York, 1953.

BIBLIOGRAPHY

Appleyard, R., *Sir Charles Parsons*. Constable, London, 1933.

Armstrong, W. G., 'Water Pressure Machinery', *Proc. I.M.E.*, 1858, p. 126, and 1868, p. 21.

Barwell, F. T., *Lubrication of Bearings*. Butterworth's Scientific Pubns., London, 1956.

Buchanan, R., *A Treatise on Mill Work*. London, 1841.

Chalmers, T. W., *Historic Researches. Chapters in the History of Physical and Chemical Discovery*. Morgan Bros., London, 1949.

Clerk, Dugald, *The Gas, Petrol and Oil Engine*, Longmans Green, London, 1916, 2 vols.

Crowther, J. G., *Discoveries and Inventions of the Twentieth Century*. Routledge and Kegan Paul, London, 1955.

Dickinson, H. W., *A Short History of the Steam Engine*. C.U.P., 1938.

Donkin, B., *Gas, Oil and Air Engines*. Charles Griffin and Co. Ltd., London, 1911.

Ewing, J. A., *The Steam Engine and other Heat Engines*, 3rd edn., C.U.P., 1920.

Ewing, J. A., *The Mechanical Production of Cold*. C.U.P., 1908.

Finkelstein, T., 'Air Engines', *The Engineer*. April 1959.

Forbes, R. J., *Man the Maker*. Constable, London, 1950.

Gibson, A. H., *Osborne Reynolds* (monograph for the British Council). Longmans Green, London, 1946.

Kastner, L. J., 'Century in the History of Reciprocating Internal Combustion Engine', *Proc. I.M.E.*, vol. 169, 1955, p. 303.

Kennedy, A. B. W., *The Mechanics of Machinery*. Macmillan, London, 1886.

Macmillan, R. H., *Automation, Friend or Foe?* C.U.P., 1956.

Parsons, G. L. (editor), *Scientific Papers of Sir Charles Parsons*. C.U.P., 1933.

Parsons, R. H., *The Development of the Parsons Steam Turbine*. Constable, London, 1936.

Rankine, W. J. M., *A Manual of the Steam Engine and other Prime Movers*. Charles Griffin and Co. Ltd., London, 1859.

Rankine, W. J. M., *A Manual of Applied Mechanics*. Charles Griffin and Co. Ltd., London, 1858.

Rayleigh, Lord (J. W. Strutt), *The Theory of Sound*. 1877.

Reuleaux, F., *The Constructor*, trans. by H. H. Suplee, 1893.

Reuleaux, F., *Kinematics of Machinery*, trans. by A. B. W. Kennedy. Macmillan, London, 1876.

Richards, G. T., *The History and Development of Typewriters*. Sc. Mus. Pubn., H.M. Stationery Office, London, 1948.

Roe, J. W., *English and American Tool Builders*. Yale U.P., 1916.

Rolt, F. H., *Gauges and Fine Measurements*, 2 vols. Macmillan, London, 1929.

Rolt, L. T. C., *Isambard Kingdom Brunel*. Longmans Green, London,

Rouse, H. and Ince, S., *History of Hydraulics*. Iowa Inst. Hyd. Res., State University, Iowa, 1957.

Singer, C. (et alii), *A History of Technology*, vol. v. O.U.P., 1958.

Stone, W., 'Mechanism of Lubrication'; pamphlet reprinted with corrections and additions, from *Industrial Australian and Mining Standard*. Melbourne, 1922.

Timoshenko, S. P., *History of Strength of Materials*. McGraw Hill, London, 1953.

Todhunter, I., and Pearson, K., *History of the Theory of Elasticity*, 2 vols. C.U.P., 1886 and 1893.

Usher, A. P., *A History of Mechanical Inventions*. Harvard U.P., 1954.

Usher, A. P., '*Industrialization of Modern Britain*', Technology and Culture. Wayne State U.P., Spring 1960, p. 109.

Woodbury, R. S., *History of the Gear Cutting Machine*. M.I.T.P., 1958.

Woodbury, R. S., *History of the Grinding Machine*. M.I.T.P., 1959.

Woodbury, R. S., *History of the Milling Machine*. M.I.T.P., 1960.

CHAPTER VIII

The Age of Mechanical Road and Air Transport, 1900–1940

The history of mechanical engineering during the first forty years of this century was affected profoundly by the First World War (1914–18) which provided the greatest stimulus to mechanical invention that the world had known. Industrial resources of every kind were devoted to experiments and trials of new mechanisms for destruction, on a scale that was without precedent in all the countries involved in the struggle. When the fighting ceased the results of this work remained as new knowledge concerning the construction and operation of machinery, and were ready to be applied to the making of machines for peaceful purposes.

The rapidity of technical progress during the First World War created an awareness among scientists and engineers that an industrial country should devote substantial resources to research and development in times of peace in order to keep pace with the rest of the world. Thus the National Physical Laboratory in Britain and the National Bureau of Standards in the United States, both of which had been founded at the beginning of the century (1900 and 1901) were greatly enlarged after the war as were similar establishments devoted to scientific research elsewhere. In Britain, a new department of the Government was set up and styled the Department of Scientific and Industrial Research (D.S.I.R.) to provide encouragement of every kind for the promotion of scientific research; by 1940 more than twenty co-operative research organizations had been established in Britain for investigating technical problems of importance to particular industries. In addition, many industrial firms and government agencies established their own research organizations and funds were made available to the universities to engage in research work on a substantial scale. The day of the lone inventor with his slender resources was gradually drawing to a close.

In the United States the speed of change was much greater. This was evidenced by more lavish expenditure on research, by a greater output of engineering graduates (in proportion to the population) by a larger annual increase of productivity per man in industry and by the more rapid development of new machines such as aircraft. It was in America that the conquest of the air, in heavier than air machines, was first achieved by the Wright brothers in 1903 and it was in that country thirty years later that the modern airliner was born.[1] Three machines introduced in quick succession at that time —the Boeing 307 (Fig. 281), the Douglas DC1 and the Lockheed Electra—possessed features that were to become standardized for airliners up to the present day.[2]

Some of the most important scientific discoveries in aeronautics were made in Germany, where the rate of scientific advance in many branches of technology was ahead of the rest of the world at the beginning of the century and continued to be so until the end of this period despite her defeat in the First World War. In building aeroplanes and airships, in devising machinery and equipment for high-pressure chemical processes, in making optical and other instruments, machine tools, and in developing welding techniques, Germany was in the forefront of engineering progress. Another country which excelled in producing machines with meticulous attention to details of design was Switzerland. Famed for many years for the excellence of their clockwork and watches, the Swiss applied similar standards of analysis and workmanship to the production of machine tools, engines, turbines and locomotives. One of the best examples of the new trends in machine design introduced at the beginning of this century through the application of scientific methods of analysis, is to be found in the textbook *Die Dampfturbin* by the Swiss professor Aurel Stodola.[3] His book, which has been translated into many languages, is one of the most outstanding textbooks of the period, and therein the author discusses not only the thermodynamic issues involved in turbine design but also fluid flow, vibration, stress analysis in plates, shells, and rotating discs and even thermal stresses and stress concentrations at holes and fillets—an exercise that was very advanced for machine designers in the first decade of this century.

The problems presented to engineers by using higher pressures, higher temperatures and higher speeds in machinery, made it essential that some should have been well educated in the fundamental principles of science and mathematics. In Britain engineering courses were being developed at the universities by such men as Sir Alfred Ewing at Cambridge and Professor W. E. Dalby at Imperial College in London, and in 1908, to the astonishment of many, even the University of Oxford accepted Engineering Science as a suitable subject for study by its undergraduates. The methods of studying engineering subjects at the university, devised in the previous century, were continued, but the presentation became progressively more scientific and less empirical, as one can see by comparing the textbooks of Ewing in 1908 with those of Rankine fifty years before. More use was made of mathematics, and the science of dynamics came to occupy a more and more important place in the curriculum, since it was found that many of the baffling problems produced by motion in machinery could be resolved by dynamical analysis.

In Europe and America an increasing number of engineering graduates was being produced, and in many cases they were better trained than in Britain for there were more opportunities for graduates to continue their studies to the postgraduate level by research and otherwise. Also the belief in the practical training of engineers in the workshop, as an alternative to a scientific education, continued to persist in Britain, possibly a 'hang-over' from the admiration of the achievements of the self-taught inventors of the previous period, so that the majority of mechanical engineers in this country were either self-educated or had at best received a part-time education at a technical college. Many of the Mechanics' Institutes in Britain grew during this period into technical colleges where studies were carried on—mostly in evening classes at first—of a progressively more advanced character until many of the students all over the country were taking examinations for the external degrees in engineering awarded by the University of London. After the First World war it was realized that in addition to this work and the training of craftsmen, the technical colleges could with advantage provide engineering instruction of a somewhat lower standard than that required for a university degree and such courses were standardized as National Certificate and Higher National Certificate courses, which quickly became so popular that employers gave their apprentices time off to attend in the day time in order that the students would not have to do all their studying in the evenings after working all day in the factory.

One of the consequences of the Russian Revolution in 1917 was that a drive to improve technical education in that country was begun, with the result that at the time of writing (1961) it is claimed by the U.S.S.R. that more technical engineers are being trained there than in all the rest of the world. Until recently very little information had filtered out to the free world concerning the differences between the training received there and elsewhere. However, in view of the very low standard of living that existed in Russia forty years ago and the very advanced technical achievements of the country in the last few years, it would seem that the rate of change of technical advance in the U.S.S.R. must have been greater than elsewhere and therefore may soon outpace the rest of the world.

MATERIALS

For mechanical engineers the most important material of construction during the first forty years of this century was steel. Many different varieties of steel were standardized for different purposes and each kind of steel conformed to its established standard with such certainty that its properties could be taken for granted by the designer, who was now able to specify much less surplus material in his designs than had been customary in the nineteenth century.

Uncertainties that had to be taken into account, such as the loads that might be imposed on the components of machinery during use, the magnitude of the resulting stresses or the variable quality of materials and workmanship, were all embraced in the comprehensive term 'factor of safety' often described as the engineer's 'factor of ignorance'. In practice, this meant that components had often been made much heavier, using more material than was really needed; but since the beginning of the century designers of machinery had been reducing their factors of ignorance without reducing margins of safety. Greater certainty as to what would happen to machinery in service was achieved as a result of greater skill and precision in all branches of the mechanical arts; the materials were more uniform in their composition and physical properties; components were manufactured with greater precision and assembled together with greater accuracy; possible variations of loading while in use were studied, so that they could be predicted with more certainty, and the consequent stress distribution was analysed. The leaders in this trend were the aeronautical engineers, whose ability to design machines of metal, that could lift themselves into the air and fly depended, *inter alia*, upon the correctness of the calculations of the stress-men who served the designers. The first aircraft structures at the beginning of the century were built of wood—braced with piano wire, which was the strongest steel then available. When aluminium was manufactured commercially in the second decade of the century it was used to an increasing extent on aircraft until by 1930 all-metal aeroplanes were being used on the regular air services in the U.S.A. At the same time aluminium alloys having a very high strength-to-weight ratio were being developed for use in aircraft construction.

By alloying steel with different quantities of different elements, hundreds of distinct grades of steel were produced to suit different purposes. One of the first of these was the stainless steel that has been used for cutlery since its accidental discovery in 1915. At that time the search for stainless steel was being conducted in many laboratories with the fervour that alchemists earlier had applied to the quest for the philosopher's stone. There followed in quick succession the discovery of austenitic stainless steels (the 18/8 chromium nickel steels), and later varieties that could be fusion welded because they contained small amounts of tungsten and titanium that inhibited weld decay. Early attempts to weld the austenitic steels had been followed in certain circumstances by a slow disintegration of the metal near the weld, a phenomenon that was given the name of weld decay. Other steel alloys were produced to withstand high temperatures. They were resistant to oxidation and they retained their strength and other physical properties at temperatures far above those for mild steel, which could seldom be used for temperatures above 500° F.

One of the first to observe the flow or 'creep' of metals under load at high temperatures was the English physicist Andrade, in 1910.[4] Soon afterwards engineers made numerous tests of steels at the higher temperatures at which steam boilers, super-heaters and turbines were to be operated, with the object of obtaining the greater thermal efficiencies that resulted when higher steam temperatures were used. A pioneer in this field was the engineer R. W. Bailey whose paper in 1935 on 'The Utilisation of Creep Test Data in Engineering Design'[5] set out the state of knowledge and the best practice of that time. The design of pressure vessels to withstand high pressures and high temperatures simultaneously was also the concern of engineers in the chemical industry[6, 7] when the development of high-pressure processes for the synthesis of ammonia and the hydrogenation of oil and coal spread from Germany to the United Kingdom, the United States, France and Italy immediately after the First World War. Pressures of 3,000 lb./sq. in. were usual and in some processes 15,000 lb./sq. in. were used, while the temperature in the reaction vessels was as high as 500° C. Sometimes the walls of the pressure vessels were protected from the high temperature by internal heat exchangers, and great ingenuity was displayed in devising forms of joint to withstand the enormous loads resulting from such high internal pressures—for example, the Vickers Anderson

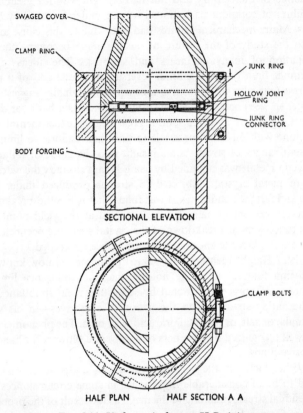

SECTIONAL ELEVATION

HALF PLAN   HALF SECTION A A

Fig. 246. Vickers Anderson H.P. joint.

joint (Fig. 246). The designers of such vessels found that they had to take account not only of stresses induced by the internal pressure but also of stresses produced by differences of temperature both radially through the thickness of the wall, and axially along the length of the vessel. Methods of reinforcing the walls of pressure vessels were developed, of which the best known were wire-winding around the circumference, and the process of 'auto-frettage' in which the walls of the vessel were initially stressed so as to produce a more desirable stress distribution throughout the wall thickness, as in a compound cylinder where one cylinder was shrunk onto another. Both methods were originally developed for strengthening the barrels of large guns.

The speeds of machinery had been increasing rapidly since the beginning of the century and this had brought with it a greater risk of failure of components by metal fatigue under repeated cycles of stress. Many mechanical failures had been due to this cause and so led to the study of the nature of metal fatigue becoming the pre-occupation of many engineers and scientists. The ideas of an endurance limit (e.g. ten million reversals of stress) and of a range of stress (e.g. plus or minus five tons/sq. in.) originally suggested by Wöhler (p. 289) had become generally accepted as a basis for design when fatigue was involved, and a great deal of experimental work was done with different materials to find their endurance limits for different ranges of stress. Some insight into the cause of failure by fatigue in metals was obtained by using X-rays to study the deformation of metal crystals subjected to stresses produced under both static and fatigue conditions. It was found that while the breakdown of the crystals into smaller grains occurred at the yield point in a static test, a similar breakdown of the crystal structure occurred in a fatigue test when the endurance limit had been exceeded.

In 1917 Haigh[8] identified the related phenomenon now known as 'corrosion fatigue' when he noticed that the endurance limit of specimens of brass was lowered by subjecting them simultaneously to alternating stress and to the action of corrosives—in his work ammonia, or salt, or hydrochloric acid were used. The phenomenon is now well recognized by designers of pumps and blowers for handling corrosive fluids.

Another factor in the behaviour of metals under stress that was found to be of considerable importance in some circumstances was the residual stress remaining in the metal as the result of the processes of manufacture. Such residual or 'locked up' stresses may be danger-ously large after cold drawing or pressing operations or after welding, and can usually be resolved by suitable heat treatment.

Studies of the behaviour of metals under stress brought about collaboration between engineers and metallurgists as a result of which engineers acquired a greater awareness of the importance of the in-dividual history of each piece of metal in a machine, from the time that the ingot was cast—through the hot forging, cold rolling, heat treatment, machining and welding—to the finished component. All these processes contribute to the suitability of the component to withstand the stress system to which it will be exposed in service.

Among the tests devised to establish the metallurgical condition of material used in machine construction were the impact tests, in which a notched bar of the material to be tested was held in a vise and then struck by a swinging pendulum which fractured the specimen at the notch and the energy absorbed by the fracture was

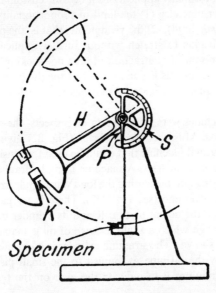

Fig. 247. Charpy impact machine.

measured. Two machines, the Charpy (Fig. 247) and the Izod were standardized; in the former a striking velocity of 17·5 ft./sec. was used and in the latter 11·5 ft./sec. The impact strength obtained from these tests was expressed in ft.-lb. and minimum values are now commonly included in specifications of metals to be used in components of machines.

The other important tests of the physical properties of material that came into general use at this time were tests of hardness. The most popular of these was the Brinell hardness test, in which a standardized hardened steel ball was forced, under a standard load, into the surface of the material to be tested. The diameter of the indentation left after the test was measured under a microscope, and used to express the hardness of the material on a scale of Brinell numbers, also used in specifications of materials. Other hardness testers, such as the Rockwell, used a spherically-tipped conical indentor of diamond and the Vickers a square pyramid diamond indentor of standard form.

The test of material of the greatest importance remained the tensile test in which a specimen is gradually stretched under load until it breaks. The stress at rupture was not the only information derived from this test; for many applications it became customary during this period to determine also (1) the limit of proportionality of stress to strain (the elastic limit), (2) the yield point; that is, the point at which the material begins to stretch without further application of load, and (3) the amount of contraction of area or 'necking' of the specimen at fracture, for this is one measure of the ductility of the metal, an important quality in material that has to be cold-worked during manufacture.

Testing machines were improved in all respects, the steelyard type of machine like Musschenbroek's lever (Fig. 82) giving place to a so-called Universal machine that could be used for tension, compression, bending or shear tests. A machine using hydraulic oil pressure both for applying and measuring the load was popularized by Amsler and Co. of Switzerland (see Fig. 248). There is no packing or cup leather between the actuating ram R and its cylinder but a very fine clearance through which a slight leakage of oil is maintained by the oil pump (not shown). The straining cylinder is connected to the load indicator F by the oil pipe A coupled to the small piston D which moves the pendulum P to indicate the load on the test piece by a pointer at F.

Universal testing machines of different types and sizes were installed in engineering laboratories in factories and colleges throughout the world so that means were available in every industrial locality for testing materials. One of the largest machines at the Berkeley division of the University of California, U.S.A., was able to exert four million pounds load. At the other extreme, one of the smallest machines in general use was the Hounsfield tensometer for miniature test pieces. It was a portable Universal testing machine able to exert a maximum load of two tons. Towards the end of this period testing machines were provided with means for adjusting the rate of strain during the test, since it had been found that this must be standardized if results were to be repeatable.

The science of experimental stress analysis was established at this time and at first its most effective method was by photo-elasticity.

It will be recalled that this method had been pioneered by Maxwell (p. 288) but in this period it was improved, first by using various transparent plastic materials that became available, and later by making it a three-dimensional method by freezing the stress in the model (of transparent material) which was later cut into thin slices

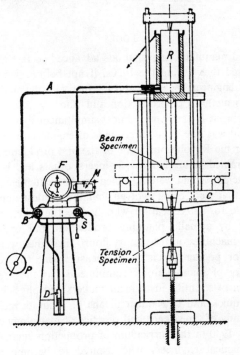

Fig. 248. Amsler testing machine.

for analysis. The two-dimensional method of photo-elastic stress analysis was popularized in the twenties and thirties by E. G. Coker[9] in London where many young research workers were trained in the method. Although it can only be used on transparent models, the particular value of the photo-elastic method is that the magnitude and direction of the stress at every point in the model can be obtained experimentally, by loading the model and then observing the colour pattern produced when polarized light is passed through it. In this way the local stresses produced at key ways, fillets, holes and welds, and in a variety of different shapes, were established.

Non-destructive testing of metals by the use of X-rays and gamma rays was introduced in the 1920s and 1930s, particularly for examining the welds in fusion-welded pressure vessels. In this method the rays were passed through the metal and photographed on the other side, the photographic print being a shadowgraph which showed up any inclusions, empty spaces or irregularities inside the weld.

TOOLS

Tools and manufacturing methods advanced so fast and became so specialized that during the 1920s, if not before, a new kind of mechanical engineer began to appear, one who devoted his whole attention to methods of production and who styled himself a production engineer. He appeared almost simultaneously in Europe and America and was one of the first kinds of engineer to be trained in Russia after the Revolution. Some of the first production engineers had been trained as mechanical engineers, others had been trained in other disciplines and were attracted by the excitement of the new techniques that were being evolved, but the greatest numbers were men of little formal education who had learned their production engineering by actually planning production and operating and supervising machines on the factory floor. These men were aware of the need for greater precision of measurement to secure interchangeability of manufactured components, so they secured the co-operation of scientific instrument makers and physicists in their search for quick and convenient means of accurate measurement. Thus the study of the science and practice of metrology came to be included by 1940 in the curriculum of production engineers and of those mechanical engineers who aspired to become production engineers and managers. The extent of this change may be observed from the successive adoption of working to standardized limits and fits, the use of 'go' and 'no go' gauges, end-measuring machines, gauge blocks, optical flats, the sine bar (Fig. 249), auto-collimators, (Fig. 250), comparitors, gear-measuring machines, the profile projector, the universal measuring machine, the workshop microscope, and the profilometer. In these advances the lead was taken in Britain by the National Physical Laboratory (N.P.L.) in London, and in the United States by the National Bureau of Standards in Washington. One of the pioneers of fine measurement in Britain, F. H. Rolt[10] of the Metrology Division of the N.P.L., performed a great service to production engineers by the publication in 1929 of his textbook

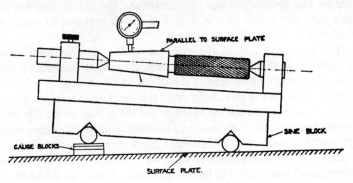

Fig. 249. Sine bar.

(in two volumes) on gauges and fine measurements. These volumes have remained a standard work of reference on the subject to this day.

An outstanding advance in metrology was made in 1908 by the Swedish engineer C. E. Johansson when he introduced the block and slip gauge system. A typical set of block gauges when they were first on sale consisted of 81 small rectangular blocks of hardened steel ranging in length from 0.05 to 4 in. Each block was ground and lapped on two opposite surfaces to such a high degree of accuracy that any two gauges could be joined together to make a thicker gauge by 'wringing' them together—that is, by first cleaning them in spirit and then bringing their faces together by hand with a combined

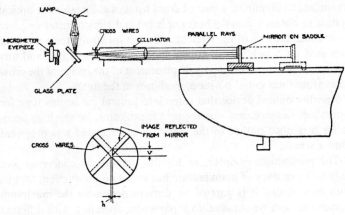

Fig. 250. Auto-collimator.

sliding and twisting motion. By combining the blocks together in this way, any gauge lengths desired could be obtained in steps of 0·0001 in.

For measuring angles the principle of the sine-bar came into general use in the tool room. This method involved determining the sine of the angle by measuring the height which one end of a bar—the sine bar—had to be raised from a surface plate to form the angle to be measured between the sine bar and the surface plate. Fig. 249 shows this method being used to measure the angle of a taper plug gauge. The gauge was held between centres mounted on the sine bar, one end of which was raised by slip gauges until the upper side of the taper gauge was parallel to the surface plate. The sine of the angle of the taper could then be calculated from the thickness of the slip gauges divided by the centre distance between the Vee grooves on the sine bar. The angle in degrees was then read off from a table of sines of angles.

Many similar adaptations of simple geometrical principles are now used regularly for measurements in the workshop and the tool room.

The application of optical methods to fine measurement was a feature of workshop metrology during this period. An ingenious device of this kind was the auto-collimator used for levelling and for tests of alignment and straightness. The principle is shown diagrammatically in Fig. 250 where the elements of the auto-collimator are shown for determining the straightness of a machine bed. The auto-collimator was a telescope containing a small lamp and a glass plate set at an angle (so that light from the lamp was reflected from its surface to illuminate a pair of cross wires) a collimator (or optical system to obtain a parallel beam of light) and a micrometer eyepiece in line with the axis of the telescope. To test for straightness a mirror mounted on a saddle was moved along the machine bed and any inaccuracies were shown up by displacement of the image of the cross wires from their initial position, as shown in the figure.

Another optical device that came into general use at this time for workshop measurement was the profile projector, in which an accurately magnified outline of the object to be measured was projected onto a screen.

The production engineer, as his title implies, is concerned not only with accuracy of manufacture but also with production, that is, with output, and it is part of his duty to determine the maximum output that can be obtained in a particular operation with different types and arrangements of machines, tools, and operators. The pioneer in getting more production from machine tools was F. W. Taylor, an American who has been styled 'the father of scientific management'. Among many contributions to higher productivity, his discovery of a high-speed tool steel that would retain its hardness while red-hot was acclaimed as the outstanding development in metal-working machinery at the Paris World Fair in 1900.[11] Taylor later described[12] the events that led to his search for a better tool steel. He had already invented an automatic grinder, and made improvements to chucks, boring mills, forging and tool-feeding mechanisms. The trail he blazed led to a number of distinct and separate advances. First, machine tools of all kinds began to be made much heavier and stiffer to take advantage of the heavier cuts that were now practicable with high-speed cutting tools. Secondly, the heavier cuts required more power at the machine spindle and this became even more evident when still better cutting tools of tungsten carbide were produced in the 1920s. The move to provide a separate electric motor drive for each machine tool, which began before 1900, had gone so far by 1940 that very few machine tools were then sold to be driven by belt from a countershaft. Usually, the electric motor incorporated in the machine provided independent drives for spindles and feed screws.

The design of machine tools entered upon a new phase after the presentation and general acceptance of Professor Schlesinger's publications on machine-tool tests and alignments.[13] Schlesinger put forward proposals, first in Germany and later in Britain, that when machine tools were being tested for acceptance by the purchaser, the tests should be standardized and include not only tests of the speed and accuracy of work done on the machines but also of the rigidity of the machine tool, by measuring its deflections when operating at maximum output.

Taylor did much to introduce the scientific approach to all production problems in the workshop. One of its consequences was the displacement of rule-of-thumb methods in hardening and tempering of tools by the application of up-to-date knowledge of physical metallurgy which spread until there came to be in many workshops a very close liaison between the production engineer and the metallurgist.

Another field for close co-operation appeared when welding was adopted as a means of manufacturing components, pressure vessels, and machine parts previously made as castings. Many serious mistakes were made by mis-applying welding in the first thirty years of this century, but gradually the conditions for obtaining satisfactory welds came to be appreciated, so that by 1940 it was the accepted method of manufacture for many articles. One condition required in the making of certain welded pressure vessels was that the whole vessel should be normalized or stress-relieved in a furnace after manufacture. The testing of such vessels by hydraulic pressure was also standardized and many tests were devised for determining the skill of manual welders and automatic welding machines by welding sample test pieces for destruction, by tensile tests, twisting tests, bend tests, and by notching and breaking the weld for visual examination. None of these tests produced certain evidence of the condition of the welds in the finished work, but a number of non-destructive tests were devised for doing this. The most important was the use of X-rays and gamma rays which were used to photograph all the welds in a finished article to show up any porosity or inclusions or other defects in the welding.

Much of the pioneer work in developing techniques for welding was done in the United States where standard codes for the construction of welded unfired pressure vessels included X-ray testing as early as the 1930s. Automatic machines for welding by electric arc with specially prepared electrodes were developed in the United States where before 1930 thousands of miles of piping had been welded for bringing natural gas from the oilfields of the south to the industrial centres of the north. In many parts of the world the oil companies, with their need for long pipe lines, did a great deal to pioneer the use of welding. The oxyacetylene process for welding steel both manually and by automatic machines was mainly developed in France, where great skill was acquired and excellent welds made by this process. Its popularity quickly spread to other European countries, but for the welding of steel it gave place in most applications to the electric arc process before 1940.

The oxygen cutting of steel which came into general use in this period is perhaps the most spectacular process to be seen in the workshop. An oxy-gas flame is impinged on the surface of the steel where it is to be cut, and before a melting temperature is reached, the iron in the steel begins to burn in the oxygen atmosphere provided by the torch, forming iron oxide, which flies off in a shower of sparks. The process was used for cutting off the tops of ingots in steelworks for it is suitable for cutting great thicknesses; it is also faster than other methods, and in the workshop it can be used to cut steel with considerable accuracy—for example, to one sixteenth of an inch or less.

It was one of the early workshop processes to be made fully automatic, by fastening the torch to a pantograph mechanism moved by a cam passing round the groove in a template, the groove having the same contour as the article being cut. In later machines a drawing took the place of the template and the pointer which replaced the cam was kept over the lines on the drawing by an electric lamp and photoelectric cell. The method of copying the contour of a template used in the oxy-cutting machine in two dimensions was applied in three dimensions in the Keller die-sinking milling machines introduced in the 1920s for profiling the large dies used for pressing the steel bodies of automobiles. In this machine a tracer cam moved with a light pressure over the surface of a wooden pattern and its movements were fed into a servo-mechanism which controlled the various feeds of the milling machine. The feeds and drives on these machines were hydraulic, a convenient method for providing infinite variation of speed or feed.

Two of the most significant developments in machine tools were the greater use of hydraulic drives and controls on machine tools, introduced by Brown and Sharpe in the United States in 1902, and the unit construction of machine tools in which standardized units such as headstocks, tables, saddles and slide ways were put together to form special machine tools for performing a number of operations simultaneously, a system particularly useful in the automobile industry.

A new machine tool made its appearance in the tool room during this time. It was the precision jig boring machine, developed from the machines used in Switzerland for the precision drilling of the jigs used in watchmaking. The special feature of the jig borer was that all the screws for positioning the work and the head were made with the accuracy of micrometer screws and the tables and slides could be moved with corresponding accuracy. Jig boring machines capable of taking very large castings on the table, such as gear-box casings for turbines, were supplied with guarantees that holes could be positioned and bored with an accuracy, from one hole to the next, of one ten-thousandth of an inch or less. To preserve such accuracy the machines were usually housed in an air-conditioned room, the air being filtered and maintained at a constant temperature and humidity. Jig boring machines of slightly different designs were developed in Switzerland, Germany, the United States and Britain.

An interesting innovation to the gear-hobbing machine was made by C. A. Parsons—the inventor of the steam turbine—when he realized that inaccuracies in the hob and the machine drive could be spread around the periphery of the blank being cut, by mounting it on a table C which was itself driven by another B, A being a driving sprocket (*see* Fig. 251) so that there was a 'creep' of about 1 per cent between the two tables. By this means more accurate and quieter gears were obtained with a higher mechanical efficiency than before.

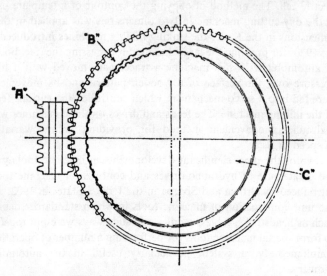

Fig. 251. Parsons's creeping table.

A completely new principle was introduced into the workshop around the turn of the century with the invention of the magnetic chuck. This device, which consists of a set of permanent magnets whose flux path passes through the work on the chuck and can be broken by turning a key which interrupts the flux path, found immediate application for surface grinding, particularly of strip material, which is difficult to hold. Recently similar magnets have been marketed for holding steel plates in position prior to welding, and for holding clock gauges and scribing blocks on the 'marking out' table.

The most important innovation in grinding was the introduction of centreless grinding in 1915. L. R. Heim in the U.S.A. patented its essential features, which were the blade to support the work slightly above the centre line of the wheels (*see* Fig. 252) and the regulating wheel rotating much slower than the grinding wheel and in the oppo-

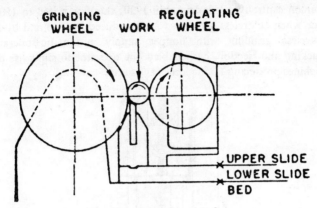

Fig. 252. Centreless grinding.

site direction. Earlier attempts at centreless grinding resulted in the work being 'lobed' (*see* Fig. 253), that is, having a constant diameter but being out of round. This was corrected by fixing the work blade in such a position that the work was above the centre-line of the two wheels. Centreless grinding can easily be made fully automatic so

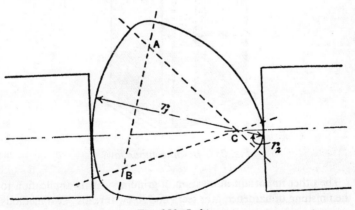

Fig. 253. Lobing.

that highly-skilled operators are not needed. It is also capable of great accuracy and fine finish but its greatest merit is the extremely high production rates that can be achieved with consequent low operating costs. Woodbury quotes[14] output figures for the grinding of hardened steel valve tappets which were done at the rate of 90 per hour on centre-type grinders until 1920; this was raised to 150 per hour when centreless grinding was introduced in 1923 and by 1929 centreless grinding with hopper supply, automatic operation, chucking and ejection, had enabled one operator to supervise three machines producing 1,350 pieces per hour.

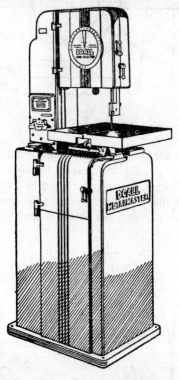

Fig. 254. Do-all contour saw.

The other important innovation in grinding was its application to the finishing of hardened gear teeth and screw threads. Two methods were found for doing this—the first by constantly trueing the face of the grinding wheel to the form required, with a diamond point, and the second by crush forming, that is by impressing the desired form on the surface of the grinding wheel by rotating it while it was pressed up against a hardened steel roll on which the desired form had been produced. The grinding process was applied on an increasing scale for production work, particularly for the motor industry where high rates of output and low cost of manufacture were so important. Special machines were designed for grinding cylinder bores, camshafts, crankpins and practically all the working surfaces of the components of the automobile. As an example of rates of output Woodbury quotes the grinding of piston rings in a disc machine, both sides of the ring being ground simultaneously, removing one thousandth of an inch of metal with a tolerance of two-tenths and finishing parallel to one ten-thousandth of an inch, the rate of production from a single machine being 10,000 piston rings per hour.

A new machine tool that proved to have many uses was the metal cutting band-saw and continuous filing machine introduced in 1933 in the United States under the name of the 'DO-ALL' machine (Fig. 254). It was provided with a tiltable table for sawing or filing at precise angles and with variable speeds and a set of saws and files for cutting a wide variety of different materials. The files were made up in short lengths fixed to a flexible steel band.

MACHINES

There was a revolution in transport by land in all civilized countries during the first forty years of this century. It was more evident in the U.S.A. than elsewhere, because there the pace of change was greatest and that country was assuming the industrial leadership of the world. People came to depend more and more on motor transport by road. By 1940 there were 25 million automobiles in the United States and fifty thousand communities there depended entirely on motor transport to move themselves and their supplies and products. There were then more than 4 million trucks and 130,000 buses.

Meanwhile the railroads, which reached their zenith about 1920, had started to decline. From 1921 to 1938 the number of passengers carried on the Class 1 railroads in the United States was more than halved, in spite of expensive improvements to provide more comfort for passengers. These improvements entailed providing extra space and weight; for example, whereas the weight of the carriage was 1,000 lb. per passenger in 1900, by 1920 this had increased to 2,300 and by 1930 to 2,500, the last increase being due to the provision of air-conditioning equipment. During the 1930s lightweight high-speed trains were introduced both in the United States and in Europe. Aluminium alloys, high-tensile and stainless steels featured largely in the construction of these trains which operated at speeds of 100 miles/hr. or even more. Diesel-electric locomotives were generally used because they had many advantages—rapid acceleration due to high tractive power at low speeds, lower maintenance and fuel costs, and lower rail stresses.

One of the factors that caused the decline of the railways was the availability of cheap mass-produced automobiles. Its pioneer in the United States was Henry Ford who, between 1908 and 1913, revolutionized the manufacture and assembly of his famous Model T

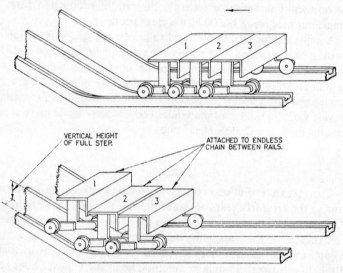

Fig. 255. Principle of escalator.

automobile (Fig. 282), which continued to be the only model he produced up to 1927, the total number built during this time exceeding 15 million vehicles. The idea of conveyor belt assembly was not new, but Ford took it up and was the first to apply it to the automobile and the first to do so on such a large scale. Many others followed the same method which was found to produce large savings because it involved better organization of the flow of materials, so that there was less time spent in handling, and above all the factory workers could never waste time, since 'the iron foreman', as the conveyor belt was called, kept them perpetually 'on the job'.

An interesting variation of the belt conveyor that came into general use about the beginning of the century was the escalator (Fig. 255) which was found to be a most effective way of moving large numbers of people up or down stairs in shops and in approaches to underground tube railways. At first vertical lifts or elevators were used, but owing to its inherent safety, peak load performance, and continuous automatic operation the escalator came to be preferred for short lifts.

The adoption of conveyor belt assembly was an important step towards the automatic factory and led to the use of automatic devices in manufacture. For example, in 1924 Archdale in England made for Morris the first automatic transfer and clamping machine for machining cylinder blocks for automobiles, and before 1930 the A. O. Smith Corporation of Milwaukee, Wisconsin,* were operating an automatic machine for making chassis frames for automobiles at the rate of one frame every six seconds. The machine comprised nine separate units linked by automatic handling gear which inspected the strip steel as it entered, rejected any that was unsatisfactory, cleaned, folded, drilled and cold riveted the rest, and washed, dipped and stove-enamelled each completed chassis before it was dispatched. This was the nearest approach to an automatic factory then operating, for the material was untouched by hand throughout, and its progress from one stage to the next was controlled automatically.

A mechanical toy that had been known for centuries—the gyroscope (Fig. 256)—was applied to certain automatic controls at the beginning of the century. It is essentially a heavy flywheel which is rotated at high speed, thereby acquiring gyroscopic inertia that enables it to resist any attempt to disturb it from its axis of rotation. The first to apply this device successfully to damping the pitch and roll of a ship was Schlick in Germany in 1903. He mounted a heavy flywheel in a frame with its axis of spin vertical and to allow for alteration of the course of the ship, the frame was free to turn round a horizontal axis athwartships. Soon after this Dr. Elmer Sperry in the United States became interested in the stabilizing properties of the gyroscope and in 1913 began fitting them in United States warships, and merchant vessels. He claimed that the roll of a 26,000-ton vessel could be virtually eliminated by a stabilizer weighing 180 tons.[15] The cost, weight and space of these large stabilizers soon led to the substitution of a small gyroscope as the sensing element, to control the movement of retractable fins projecting from the sides of the ship as an anti-roll device. The gyroscope was also the essential element in the gyrocompass,[16] the gyro pilot, and of the aeroplane stabilizer and aircraft automatic pilot.

Automatic control had begun to be established in the process

* Visited by the author in 1930.

Fig. 256. Gyroscope.

industries, of which power production is the closest to mechanical engineering. In 1912 the Crosby feed-water regulator became popular for automatic control of the water level in steam boilers. This device (Fig. 257) depends upon heat transfer through a metal diaphragm for its operation. The diaphragm was contained in a small metal bulb outside the boiler, at the height where the water level in the boiler was to be maintained, the space above the diaphragm and the piping connected to the feed-water control valve being filled with distilled water, and the space below the diaphragm being connected to the steam and water connections to the boiler. If the water level in the boiler rose, water from the boiler cooled the diaphragm, the distilled water contracted and closed the feed-water valve; conversely, if the water level fell, steam from the boiler warmed the diaphragm and the feed-water valve opened. This contrivance was very sensitive and when steam was being drawn from the boiler the valve was continually oscillating. This is but one example of many process controls which came into use in this period.

Many different types of apparatus were used for the automatic control of the combustion in steam boilers. Test results of a number were given in 1934 in a paper by J. L. Hodgson and L. L. Robinson,[17] where no less than eleven systems were examined and discussed. These were known as the Arca, Carrick, Enco, Roucka, Smoot, Hagan, Askania, Leeds and Northrup, Siemens, Bailey and Kent systems and there were others existing at the time. The authors

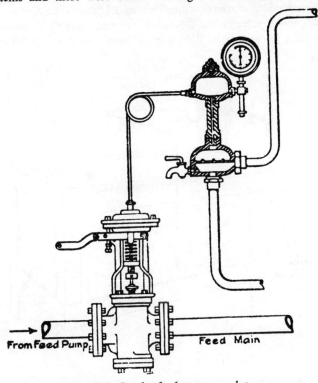

Fig. 257. Crosby feed-water regulator.

showed that between 1907 and 1931 the maximum boiler efficiency on large power stations rose from 70 per cent to 89 per cent, largely because of automatic combustion control, which helped in smoke prevention apart from fuel saving. Most of the systems based their initial point of control on the steam demand through the steam pressure, thus—together with minor subsidiary controls, such as CO_2 content in flue gases—controlling the quantity of fuel and air required for each operating condition.

For speed control the power of the centrifugal governor was increased by introducing an oil relay system, oil pressure being supplied by a separate gear pump so that all the centrifugal governor was called upon to do was to open or close a piston valve in the oil-pressure relay system that operated the main steam valve. This arrangement was first used by Parsons in 1917 for governing an 8,500 kW turbo-generator supplied at Sheffield, England.[18]

The same idea of using another fluid under pressure to operate the main control valve, while the sensing element merely operates a relay valve in the pressure fluid line, had been used about 1900 by the

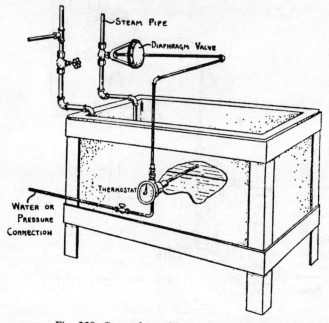

Fig. 258. Steam-heated hot-water regulator.

Cambridge Instrument Company to control the temperature of a steam-heated hot-water system (Fig. 258). Here the sensing element or thermostat was of the rod and tube type in which a central rod of Invar (a nickel-steel alloy having a very low thermal expansion), was surrounded by a brass tube immersed in the water to be heated, and the difference in length of the two, as the temperature changed, operated the relay piston valve in the pressure-water supply and this in turn operated the steam valve by means of a diaphragm.

Bellows made of thin metal sheet came into use as thermostat elements about 1914 when the United States Kelvinator Company used them to operate an on-off electric switch to make a refrigerator automatic. (Metal bellows had been used as the essential element of the aneroid barometer since its invention by L. Vidie in France in 1844.) Two other thermostats that are now commonplace were introduced in the 1920s, Payne's thermostat (Fig. 259) for automobile

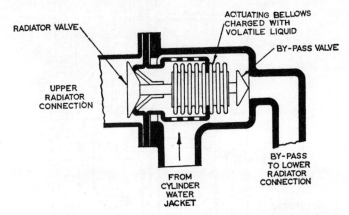

PAYNE'S THERMOSTAT FOR AUTOMOBILE COOLING SYSTEM. 1925

Fig. 259. Payne's thermostat.

cooling systems used a metallic bellows charged with volatile liquid, to operate both the main valve to the radiator and a by-pass valve. The other was the thermostat used on the domestic gas oven, which was a rod and tube type but incorporated a small by-pass connection so that the gas was not entirely cut off from the oven when the thermostat valve was closed.

The theory of automatic control was advanced considerably in 1932 when H. Nyquist in the United States, in a paper entitled 'The Regeneration Theory'[19] started to consider the effect of superimposing regularly fluctuating disturbances on a control system at different frequencies, to find under what conditions self-sustaining oscillations could arise. This led to making frequency response tests of each component of a control system, so that the characteristics of different systems could be pre-determined and the desired performance obtained. Important studies of the effects of time lag on the operation of control systems were published in 1936 by three British collaborators, A. Callendar, D. R. Hartree and A. Porter.[20]

Some important advances were made in this period in the study of vibration in machinery. It became essential for the designer to know,

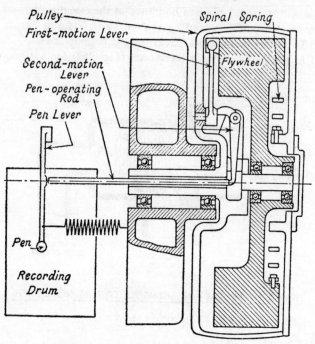

Fig. 260. Geiger torsiograph.

before his machine was built, the magnitude and frequency of the pulsating stresses induced in the components during operation, particularly when parts were reciprocated at high speeds or when large out-of-balance masses were subjected to the repetition of a cycle of events. Some examples where these considerations were important were large marine engines, high-speed piston engines for aircraft, and rolling mill machinery, though there were many others where the effects of vibration of machinery placed a limit on operation. Vibrations may be linear—that is, up and down, sideways, or endways—or angular, that is, a shaft or member may be periodically twisted and untwisted as it rotates, and the control of such vibration, known as torsional vibration, has provided excellent opportunities for the successful application of mathematical analysis to its solution. Such studies have shown that sometimes it is enough merely to alter certain dimensions of the machinery in order that the natural frequencies of oscillation be altered; sometimes it is necessary to introduce a vibration damper or vibration absorber that will dis-

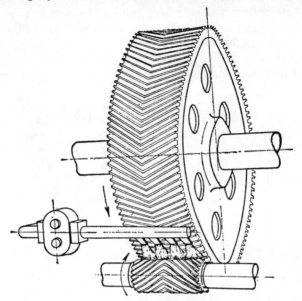

Fig. 261. Helical gearing. (*single reduction*)

sipate the vibration energy as conditions of resonance are approached. Many different types of vibration damper were devised, one of the earliest for torsional vibrations being that of F. W. Lanchester who, on one of his automobiles interposed in the drive from the engine, a flywheel driven through spring-loaded friction surfaces from the crankshaft, so arranged that at a certain frequency some slipping would occur at every cycle thus dissipating some of the energy of vibration.

The practical study of vibration requires that some instrument should be available for observing it. The device that came into general use by 1930 for the observation of torsional vibrations was the Geiger torsiograph (Fig. 260) which was driven by a light cotton belt from the shaft whose angular vibrations were to be measured. The vibrations were recorded continuously by the pen on a moving strip of paper on the recording drum.

Two of the major achievements in mechanical power transmission were the introduction of double-reduction helical gearing (Fig. 261) for transmitting very large powers, as from the steam turbine to the propeller on a large ship, and the use of multiple-Vee belt drives, usually for driving small machines of about 5 to 50 h.p. from individual electric motors, which gave greater safety, and more daylight, in factories and workshops as line shafting, pulleys and miles of belting became almost a thing of the past by 1940.

An interesting machine that was technically sound but a commercial failure in a number of different countries was the swashplate

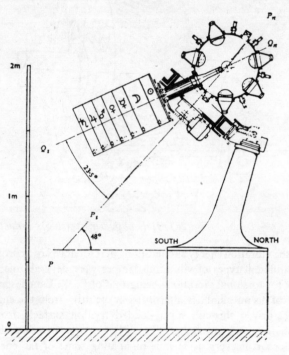

Fig. 262. Planetarium.

crankless engine promoted by A. G. M. Michell, the successful inventor of the tilting pad thrust bearing. Michell built* a crankless automobile engine in Melbourne, Australia in 1911, and during the next thirty years advocated this form of construction for pumps, compressors, automobile engines and aircraft but with very little success. There was an eighteen-cylinder aircraft engine of this kind being built in Cleveland, Ohio, in 1930 but it never got beyond the prototype stage.

Another interesting machine that achieved success and popularity was the projection planetarium (Fig. 262), built first at Munich in

* Communicated to the author by A. G. M. Michell in 1940.

1923 by Professor Walther Bauersfeld.[21] It was a small optical projector placed within a very large hemispherical dome on the inside surface of which the motion of the sun, moon, planets and stars could be displayed to an audience seated within the dome, by imparting suitable regular motions to the projector. This apparatus proved extremely popular both for instruction and entertainment and similar planetaria have since been built in Berlin, Chicago, Milan, Moscow, London, New York, Los Angeles and indeed in many large cities all over the world.

FLUID MACHINES

The most important fluid machine in the first half of this century was the aeroplane. The present chapter covers the period from the first aeroplane flight by the Wright brothers at Kitty Hawk in the U.S.A. in 1903 to the production in quantity of the Spitfire fighter aircraft at the beginning of the Second World War in 1939. During the thirty-six years that separate these two achievements the growth of human knowledge about fluid flow was prodigious. Each year new discoveries were made that enabled engineers to design aircraft that were safer and faster than their predecessors, and a great many of these discoveries were concerned with the motion of solid bodies through a fluid—in this case air. The whole body of knowledge that had been accumulated about hydraulics was of little direct use to the aeronautical engineer and indeed the conventionally trained mechanical engineer with his dependence on empirical formulae and factors of safety, was often found unable to adapt himself to meet the exacting requirements for the successful design and operation of aircraft.

Until the first war in 1914 flying was not taken seriously except by a few enthusiasts, but after 1918 it was generally appreciated that the construction of aircraft for both military and civilian uses was likely to provide manufacturing industry with an outlet of considerable size and economic value. Accordingly research establishments were set up to devote themselves to aeronautical studies, departments of aeronautics were established in universities and technical institutions and new journals were started where aeronautical problems could be discussed, and annual exhibitions, rallies and contests were held to foster interest in the development of new machines.

The importance of the Wright brothers' achievement in flying a power-driven machine that was heavier than the air can hardly be overestimated. For some years they had been studying what

Lilienthal and others had done before them, and by 1903 they had made more than a thousand glider flights and had taught themselves to fly. They even made wind-tunnel tests on models and designed and built the engine of their first aeroplane themselves because they could not obtain one with a sufficiently high power/weight ratio. It was a four-cylinder in-line water-cooled petrol motor of 4-in.

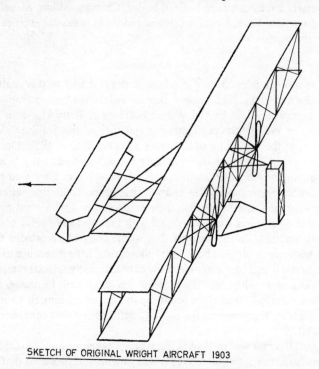

SKETCH OF ORIGINAL WRIGHT AIRCRAFT 1903

Fig. 263. Wright brothers' aeroplane.

bore and stroke, designed to develop 12 h.p. at 900 rev./min. Its weight was 180 lb. and the output on test was 16 h.p. The weight of the aircraft was 750 lb. and the area of the wings 150 sq. ft., the wing span being 40 ft. The two propellers rotated in opposite directions and were chain driven from the engine at the rear. The machine (Fig. 263) was launched from a trolley on rails. The speed of the first flight was only 35 miles/hr.! Though the Wright brothers possessed enthusiasm, courage and determination in a large measure, it is doubtful if they would have succeeded without some understanding also of the requirements of an aircraft to maintain stability in flight. A full theoretical treatment of such requirements was published in 1907 by the English engineer F. W. Lanchester in his textbook *Aerodynamics* which was followed the next year by another entitled *Aerodonetics*. These two volumes were far in advance of their time and were used as the principle works of reference in English for many years.

There followed a period of intensive experimental work to improve the performance of aircraft. Much of this was done with models and sections of aircraft wings in wind tunnels. Eventually complete air-

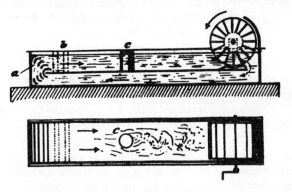

Fig. 264. Prandtl's paddle-wheel flume.

craft were tested in this way, and the data obtained about streamlining and the behaviour of the boundary layer of the fluid on solid surfaces had a profound effect on the design of all machinery concerned with the movement of fluids at high speeds. All this activity was remarkably successful. The advances made in aeronautics during the 1920s and 1930s resulted in the abandonment of hydraulics in favour of fluid mechanics as a subject of study by mechanical engineers. The originator of this new approach to the study of fluid flow was L. Prandtl who has justly been described as the founder of fluid mechanics. Prandtl, who was a mechanical engineer, made studies of fluid flow during the whole of this period, starting in 1901 in Hanover, but soon moving to Göttingen where he founded the world-famous Kaiser Wilhelm Institute for the study of fluid flow. At the outset he defined the theory of the boundary layer and he had a unique facility for appreciating the essential features of flow problems and for expressing them by simple mathematical relationships based on fundamental physical laws. He was also a skilled and resourceful experimenter who developed with his assistants new experimental techniques (*see* Fig. 264) that were copied in other laboratories all over the world. Prandtl's influence was spread not only by his published papers but also by the many brilliant collaborators who worked with him and then moved elsewhere. Among these was Theodor von Karman, originally of Budapest, who emigrated to the U.S.A. in 1930, where he continued his researches at the California Institute of Technology.[22]

The boundary layer approach to fluid mechanics was also followed in other countries, by Sir Thomas Stanton and Sir Geoffrey Taylor in England, by A. G. Eiffel who made free fall experiments in his famous tower in Paris, by Prasil in Zurich, by Riabouchinsky in Moscow and by Buckingham, Bridgeman, Durand and von Karman in the United States. Each of these and many others made their contributions to aerodynamic theory which advanced so far and so fast during these forty years. One of the most compelling reasons which led to the acceptance of the boundary layer approach by mechanical engineers was that this same method of analysis provided a basis for the study of heat transfer between flowing fluids and their boundaries as in furnaces, boilers, evaporators, condensers and heat exchangers generally. It was not until the 1920s that mechanical engineers began to apply these scientific methods to the calculation of heat transfer surfaces in their equipment and many learnt their lessons from chemical engineers who were then beginning to establish themselves as yet another kind of specialized professional engineer.

Germany took the lead in airship development,[23] inspired by Count von Zeppelin who flew one of the first of his airships over Lake Constance in 1900 and continued making improvements until his death in 1917. By 1914, twenty-five Zeppelin airships had been built. After the war the size of German airships was limited by treaty to $2\frac{1}{2}$ million cubic feet envelope until 1927, when the construction of the *Graf Zeppelin* (3·7 million cubic feet) was begun and completed the following year. Powered by five gas engines, each of 530 h.p. carried in five separate gondolas suspended from the envelope, this airship was the most successful ever built (Fig. 265). In 1929 with twenty passengers and a crew of forty the airship made a flight of 21,500 miles round the world in three weeks and during the next four years she made more than 300 cruises including 13 cruises from Germany to South America and back in the year 1933. A larger airship of similar design, the *Hindenburg*, went into regular service on the North Atlantic crossing in 1936 but the following year at the end of a successful Atlantic crossing she suddenly burst into flames while cruising round her mooring mast at Lakehurst, New Jersey, with 39 passengers and 61 crew on board.

Fig. 265. The Graf Zeppelin.

This was the culmination of a series of airship disasters. In 1921 the British airship R38 had broken into two parts while over the mouth of the Humber, with the result that 44 of the 48 persons on board perished, and in 1930 the giant airship R101 carrying many distinguished passengers and a large crew on her maiden voyage had crashed and caught fire at Beauvais, France, leaving only six survivors. The United States had fared no better, though her airships were thought to be safer since they were inflated with helium gas in place of hydrogen. The airship *Shenandoah*—very similar to the British airship R33, had broken into three parts in 1924 during squally weather over Ohio. Though fourteen lives were lost, the front portion of the ship descended as a free balloon, thus saving the lives of the Commander and 27 members of the crew. The largest American airship, the *Akron*, having a gas volume of $6\frac{1}{2}$ million cubic feet, had been lost off the coast of New Jersey in 1933 with 74 persons on board. There were no survivors and the cause of the disaster was unknown.

Large airships are inherently less safe than aeroplanes because of their greater bulk and their relative inability to resist violent involuntary movements that may be caused by weather conditions. The velocity and direction of the wind, the rays of the sun and atmospheric disturbances all have a much greater effect on the navigation of the airship than on that of the aeroplane. The airship is also slower; the fastest yet built, the *Akron*, had a maximum speed of 84 miles/hr.

The reluctance shown by many mechanical engineers to accept scientific analysis in place of empirical data was often due to their having insufficient facility in the use of mathematics. Two notable exceptions to this were the Australian A. G. M. Michell and the

American Professor A. Kingsbury, who independently invented the tilting pad thrust bearing at the beginning of the century. Both inventors applied Reynolds's hydrodynamic theory to the solution of the thrust bearing problem and they both realized that it was necessary to break up the annular area into a number of sections so that a series of self-adjusting tapered oil films could be formed around the surface of the thrust collar. The ring of tilting metal pads that bear

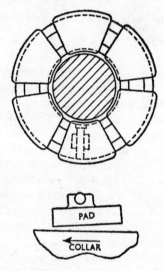

Fig. 266. Michell thrust-pad.

upon the thrust collar can be seen in Fig. 266 of the Michell thrust bearing. It has been described by Barwell,[24] in 1956, as 'the single great invention of lubrication science'. This invention, which resulted from one of the most elegant pieces of scientific analysis, was of great practical importance, for it enabled large ships to dispense with the rows of thrust collars and cooling arrangements then needed to prevent the thrust of the propeller from disembowelling the hull, and replaced them with a single bearing like that in the figure, but the invention was not exploited until Germany used it in her battleships in the First World War.

Some important developments occurred in the field of hydraulic power transmission, where a wide variety of machines were developed, some with much greater success than others. An interesting type was that where a centrifugal pump and turbine were combined together co-axially in two halves of a single casing so that the hydraulic fluid was circulated continuously round a 'taurus' formed by the casings of pump and turbine. The first of these was devised by Dr. H. Föttinger of Hamburg and Fig. 267 shows the principal parts.[25] This type of transmission was largely superseded by the more efficient

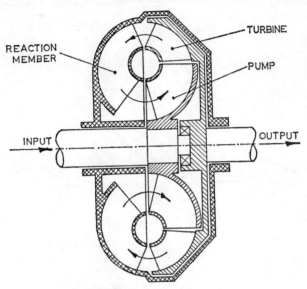

Fig. 267. Föttinger torque converter.

double helical gearing (Fig. 261) when means had been found for cutting the gears with the precision required. A somewhat similar but simpler application of the same essential idea was incorporated in the fluid coupling (Fig. 268) or fluid flywheel, developed in several countries for automobile and railcar transmissions and for some industrial purposes. Germany, the United States, Britain and Sweden all took a prominent part in these developments.

There is an important difference between the hydraulic coupling—sometimes called a fluid flywheel—and the hydraulic torque converter. The former has only two essential elements, the impeller (or pump) and the runner (or turbine), both of which rotate at nearly the same speed when the load has been taken up, and the transmission efficiency of this combination is usually about 98 per cent. In the hydraulic torque converter, however, there is a stationary reaction member through which the fluid must pass on its way from the pump to the turbine. The torque converter was designed to operate as a reduction gear with a speed ratio of perhaps 5 to 1 between pump and turbine and the transmission efficiency—usually below 90 per cent—was necessarily lower than that of the fluid flywheel because of fluid friction between the moving fluid and the stationary walls of the reaction member. Hydraulic couplings are

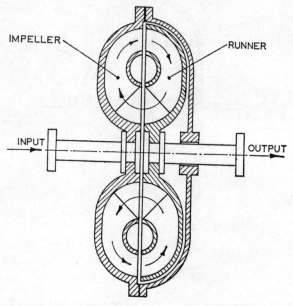

Fig. 268. Vulcan hydraulic coupling.

often used to prevent vibrations from piston engines being transmitted along the output shaft. By 1935 the largest single hydraulic couplings that had been made were for transmitting 36,000 h.p. between turbines and storage pumps at the Herdecke hydro-electric plant. Smaller units have been used extensively for marine propulsion, diesel locomotives, mine winding gear and automobiles.

A different type of hydraulic transmission—hydrostatic rather than hydrokinetic—which provided an infinitely variable speed reduction, has been obtained by using a number of reciprocating plunger pumps set in a ring about a central axis, parallel to it, and operated by a 'swash' plate or 'wobble' plate whose inclination to the axis could be varied so that the stroke of the pistons could also be varied from nothing to a maximum. This arrangement was incorporated in the Williams Janney variable-speed gear (Fig. 269). It consisted of two similar units—a pump unit and a motor unit (only one unit is shown in the figure)—connected together only by piping. Usually there were nine cylinders in both pump and motor, the angle of the swash plate of the motor unit being fixed.

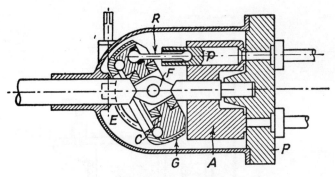

Fig. 269. Williams Janney variable speed gear.

Hydraulic drives for machinery were used increasingly at this time; for instance, for operating the ram carrying the cutting tool on a metal shaping machine; for oscillating the table of a surface grinding machine back and forth; for operating the wheel brakes on automobiles, and for the rapid closure of large valves on chemical plant. These are but a few examples of the many applications of hydraulic devices for industrial purposes.

The use of oil in very high-pressure pipe systems, say five to ten thousand pounds per square inch as in the fuel injection systems of diesel engines, led to studies being made of wave action in pipes; the principles involved have been applied to supply power by wave action through liquids for working rock drills, hydraulic riveters and other machines. A pioneer in this field was Mr. G. Constantinesco[26] whose proposals led to the making of a wave power generator for rock drills that had a reciprocating pump working at 3,000 double strokes per minute.

An important device that depended for its operation on the controlled rate of flow of both air and a liquid—petrol—was the float-feed carburettor. This apparatus (Fig. 270 is a simplified type) was invented to meter automatically to a petrol engine the amount of fuel required for any particular opening of the throttle valve. It consisted of a float chamber from which the liquid on its way to the engine passed through a venturi nozzle where air entering at high velocity met the spray of liquid petrol drawn in by the suction from the engine. By adding extra refinements working on the same principle, such as drowned jets and slow-running jets, it was possible for this apparatus to supply the right quantity of fuel under every condition of operation.

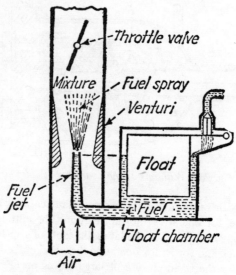

Fig. 270. Elementary jet-carburettor.

Another fluid machine to come into general use at this time was the domestic vacuum cleaner. Its essential elements were a suction fan or pump (hand-operated leather bellows were used at first), which drew the dust-laden air through a fabric bag where the dust collected —the air passing right through the fabric and into the room. Periodically the bag was removed to empty out the dust. All modern vacuum cleaners require that the place where they are used should be wired for the public electricity supply, as high-speed electrically driven fans are used to create the suction. The vacuum cleaner was invented in England by H. C. Booth[27] at the beginning of the century. The removal of particles from dust-laden air in industrial and workshop processes assumed great importance when the toxic effects of breathing dust became known.

Another interesting application of the suction fan or bellows was brought into use in the 1930s. This was the Drinker respirator or iron lung (Fig. 271) used in hospitals to maintain the breathing of patients whose chest muscles had been paralysed.[28] Invented by Professor Phillip Drinker of Harvard in the United States in 1929, it consisted of a coffin-shaped chamber in which all the body of the patient except the head was enclosed. The pressure in the chamber was reduced by only about six inches of water, about twenty times per

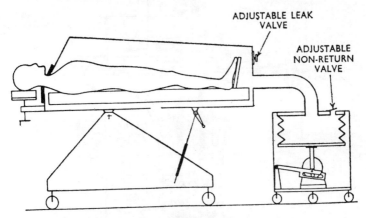

Fig. 271. Drinker respirator.

minute (or to match the rate of breathing). The Drinker respirator was the first really practical cabinet breathing machine, though the idea had been anticipated by a French physician named Woillez more than fifty years before.

The only outstanding development in water turbines was patented by Professor Kaplan in Sweden in 1913. This embodied the hydraulic version of the variable pitch propeller that also found application for aircraft propellers and for screw propellers on ships. The runner of the Kaplan turbine (Fig. 272) had two or more vanes which were turned round about their own axes under the control of the governor while the turbine was running; at the same time the angles of the inlet and outlet vanes could be adjusted without the turbine having to be stopped and by this means it was found possible to obtain a nearly constant efficiency of 85 to 90 per cent over the whole range from 30 per cent load up to full load, with water having a comparatively low head. By 1926 a Kaplan turbine of 11,000 h.p. was put into service in Sweden. The runner was 19-ft. diameter, weight $62\frac{1}{2}$ tons and it ran at $62\frac{1}{2}$ rev./min. with a water head of only 21·3 ft. The

Kaplan turbine was found most suitable for use where the head of the water supply was low or variable. For high heads of water the Pelton wheel continued to be the most suitable machine.

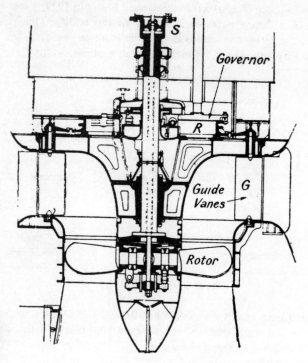

Fig. 272. Kaplan water-turbine.

HEAT ENGINES

The changes that took place in heat engines during this period had the most profound effect on material civilization in all parts of the world. This was not brought about by any major successful innovations but by many improvements made to the internal combustion engine and the steam turbine—to their details and to their accessories. Among the innovations of the period that eventually reached a measure of success, were various designs of steam boiler employing forced circulation; the Lungstrom radial-flow steam turbine with its two adjacent rotors revolving in opposite directions; the supercharging of the internal combustion engine by a high-speed rotary blower driven either mechanically from the engine or by an exhaust gas turbine; and the immense natural draught cooling towers of hyperbolic shape that now indicate on the landscape the position of large steam power stations.

As against these successful innovations, there were many that failed to reach general acceptance although they appeared to be technically sound when introduced. Among such were the crankless or swash-plate engine advocated by A. G. M. Michell, the inventor of the successful thrust bearing; Humphrey's explosion pump (Fig. 275) in which the piston, connecting rod and crankshaft of the gas engine were replaced by an oscillating column of water (a partial return to the principle of Savery's mine pump); the binary fluid power station in which mercury was vaporized in the boiler and after passing through a mercury turbine where some power was produced, was condensed in a heat exchanger where steam was raised for producing more power in steam turbines; and the steam turbine locomotive, which was a costly failure.

The internal combustion engine was made still more powerful and lighter in weight, and when mass produced in the smaller sizes, it was remarkably cheap. Refinements in every detail were introduced so that when desired it could be made extremely reliable. Thus it became accepted for the propulsion of ships, aircraft, locomotives and automobiles and reached the pinnacle of mechanical perfection in the motors constructed for automobiles and aircraft by the firm of Rolls-Royce in England whose name became a symbol of mechanical excellence throughout the world. Soon after the beginning of the century the hot-air engine almost disappeared from use. It had been very inefficient, heavy and expensive, and its place was taken by the small electric motor. On the other hand the gas engine—which was extremely efficient—began to decline in popularity after the First World War. Its high thermal efficiency was surpassed by that of the diesel engine which operated with much higher compression, and in those cases where the gas engine used cheap fuel—as for blowing blast furnaces—it was replaced by the steam turbine which required less space and less maintenance.

Many of the changes in the practice of mechanical engineering in power production resulted from the increasing use of oil in preference to coal because it was so much easier to handle, produced less smoke and dust, and was sometimes cheaper. For boiler firing on large ocean-going ships, oil had replaced coal almost completely by 1940. The same could be said of coal-burning road steamers which had had a certain vogue in the first quarter of the century. The rail-ways—particularly those in Britain—were the last stronghold of the coal-fired steam boiler and reciprocating steam engine. In the U.S.A. and continental Europe oil-burning diesel locomotives were super-seding steamers before 1940.

The development of the internal combustion engine during this period had a profound effect on all other branches of mechanical engineering. Of the four principal types of piston engine at this time the dominant position was occupied by the petrol engine, developed primarily for road vehicles and aircraft; the other three types were the large diesel engines developed for land and marine work, the high-speed diesel engine developed in the 1920s for heavy road vehicles and agriculture, and adapted in the U.S.A. and Germany for locomotives, and the gas engine which probably reached its zenith in 1914, rapidly declining thereafter. Each of these types of engine was produced in a bewildering variety both of engines and accessories.

F. W. Lanchester has explained[29] how other forms of internal combustion engine evolved from the Otto gas engine described in the last chapter. To make this engine suitable for operating a motor vehicle with petrol as the fuel, involved making radical changes and solving some fundamental problems. First, the weight had to be reduced, particularly the weight of the moving parts, for the inertia forces that they create increase as the square of the speed, so that an engine operating at a piston speed of 3,000 ft./min. creates inertia forces that are thirty-six times greater than one whose piston speed is 500 ft./min. Reductions in weight were achieved by improvement in design and by the use of materials of greater specific strength, such as alloy steels and aluminium alloys. Another characteristic of the petrol engine was that to keep the weight low, it was necessary to limit the size of individual cylinders, and to install a larger number of cylinders when more power was required; this had the further advantage of reducing the weight of the flywheel. For motor-car engines the number of cylinders went up from one to two, four, six and eight, and for aircraft engines of 1,500 h.p. eighteen and twenty-four cylinder engines were used. The multi-cylinder motor-car engines with long crankshafts introduced further complications of engine balancing and torsional vibration that were solved in due time, the latter by the inclusion of some form of torsional damper.

Many of the improvements that were made to the petrol engine and the high-speed diesel engine during this period resulted from the experimental work of Sir Harry Ricardo, who started his investigations under the inspiring leadership of Professor Hopkinson in 1904 and has continued his researches up to the present day.[30] He it was who first discovered the role of turbulence in mixing the charge and thereby ensuring rapid combustion in the engine cylinder; his contributions to the understanding of the phenomenon of detonation were also outstanding, and he was among the first to realize that detonation sets a limit to the power output and economy of the petrol engine. He was also concerned with experiments that explained why the greatest power is obtained from a richer mixture of petrol and air than the chemically-correct one, a fact that had puzzled engineers for many years, and is now recognized as being due to dissociation of the chemical compounds formed during the explosion.

By the middle of the period it was appreciated that the onset of detonation was a function both of the design of the engine (particularly the shape of the combustion space) and of the combustion characteristics of the fuel. It was discovered in 1922 by Midgely and Boyd in the United States that by the addition of minute quantities of a toxic chemical substance—tetraethyl lead—to petroleum, the tendency to detonate could be suppressed so that much higher compression ratios could be used for petrol engines.

In their efforts to try to satisfy the stringent requirements for aerial flight, engine designers adopted all kinds of stratagems. In some cases, cylinders were put in line; in the radial engine, the cylinders were radially disposed around a central crankshaft, the cylinders being fixed; in the rotary engine, with a somewhat similar arrangement, the cylinders rotated around the crankshaft. Other variations were in the method of cooling—by air, with finned cylinders, or by water or other fluids in jackets surrounding the cylinders. Great ingenuity was also displayed in the many varieties of carburettor that were devised to enable aircraft to operate not only on the level, but when climbing, banking, turning over, and flying upside down.

All engine accessories were improved to obtain more power, economy and reliability; for example, superchargers were developed to give more power for aircraft both at take-off and for flying at high altitudes; lubrication and ignition systems were improved; steels that better withstood high temperatures in the exhaust valves and others with high fatigue strength for crankshafts and connecting rods, were developed by the metallurgist.

The most important change in the development of the diesel engine was the abandonment of the air blast for introducing the fuel in a fine spray, and the substitution of the so-called solid injection system in which the liquid fuel was forced at extremely high pressure—five to six thousand pounds per square inch—through very fine holes in a nozzle in the wall of the combustion space, thus producing a fine spray without the extra weight, complication, and expense of air compressors for fuel injection. Success was achieved through an appreciation of the essential role of turbulence in the charge within the cylinder at the time of injection. The air in the cylinder was compelled to swirl very fast past the injection nozzles and 'squished' by the piston into the recesses of the combustion space so that the terrific heat generated by the sudden compression caused spontaneous explosion of the charge in that fraction of a second of time when the piston was near the top of its stroke. At first it was thought that the diesel engine could only be used for slow speeds because of the time required for the oil to burn, but studies of air movement and turbulence within the cylinder by Ricardo and others showed that the 'delay period' between the time of fuel injection and its complete combustion could be vastly diminished by altering the design of the engine so that violent turbulence was caused within the cylinder. Thus the high-speed diesel engine became practicable and before 1930 oil engines were operating at 1,000 rev./min. and by 1940 with further improvements at 2,000 to 3,000 rev./min.

Among the mechanical difficulties that had to be surmounted to make the high-speed diesel engine a success, was the fuel pump that replaced the simple carburettor of the petrol engine. This pump was required to inject precise and minute amounts of fuel at extremely high pressure with exact timing into each cylinder, fifteen hundred times a minute, on an engine rotating at 3,000 rev./min. At the same time it had to be self-lubricating, reliable, and repairable. The first man to make such pumps successfully was Bosch in Germany, but before 1940 they were being made in several industrialized countries by a number of firms.

The diesel engine can be designed to operate on either the four-stroke cycle or the two-stroke cycle. The latter can be provided for, without complication or loss of fuel to the exhaust during charging of the cylinder, since in the diesel engine only fresh air is drawn in and compressed before fuel injection. Accordingly, two-stroke diesel engines became popular both for small high-speed stationary and aero engines such as the Junkers engine in Germany, and for large marine engines such as the Doxford engine in England.[31] Both the Junkers and Doxford engines were of the opposed piston type —having two pistons in each cylinder, the pistons moving towards one another during compression and flying apart during the explosion and subsequent expansion of the charge. Fig. 273 shows a Doxford vertical marine engine of this type in which the upper piston was driven slightly out of phase with the lower one by connecting rods from the crankshaft below. The engine had what is called an end-to-end scavenge, for the ports in the cylinder were arranged so that fresh air was blown in through the admission ports at the bottom, sweeping out—or scavenging—the exhaust products through the exhaust

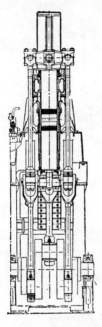

Fig. 273. Doxford engine.

ports at the top. The economy and reliability of this type of engine resulted in its adoption on a very wide scale for merchant shipping, where the shaft horse-power required was from 5,000 to 20,000 h.p.

Another innovation that has since led to great increases of specific power in internal combustion engines was the introduction of supercharging, first introduced by the Swiss engineer Dr. Buchi for marine engines, and later applied to aero-engines to boost the power output at take-off and for flying in the rarefied atmosphere encountered at very high altitudes. Supercharging was done by blowing the incoming air into the working cylinders of the internal combustion engine under pressure, the blower being either mechanically driven by the engine or independently driven by a gas turbine, utilizing part of the energy of the exhaust gases from the main engine cylinders. Such exhaust turbines were among the first gas turbines to work successfully.

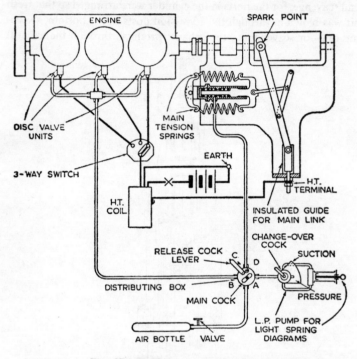

Fig. 274. Farnboro engine indicator.

The success of the experimental work carried out on internal combustion engines in this period owed much to the special instruments developed, particularly the engine indicator. The pencil-engine indicator that had sufficed for setting the valves on slow-speed steam engines in the nineteenth century was insensitive and its moving parts had too much inertia to enable it to record the rapid pressure changes occurring in the high-speed petrol and diesel engine. To meet this need optical engine indicators were devised by Hopkinson and others, and later a balanced pressure engine indicator of a novel type 'the Farnboro' (Fig. 274) was introduced in 1923 for indicating the performance of engines of aircraft in flight. This was soon found to have a much wider application, and finally, before 1940, a number of electronic engine indicators appeared, in which the bogy of the inertia of moving masses at high speed was banished for ever by replacing the masses with weightless beams of electrons. Since then electronic instruments have become indispensable for the measurement of rapid fluctuations of pressure, temperature, speed and displacement.

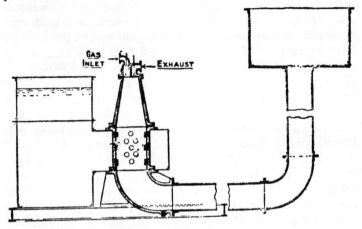

Fig. 275. Humphrey gas pump.

A novel form of gas engine that was thought to be an improvement since the ratio of expansion could be made greater than the ratio of compression, was the Humphrey gas pump in which the piston, connecting rod and crankshaft were replaced by a column of water that was caused to oscillate in a U-shaped tube by the periodic explosion of a mixture of air and gas drawn into the closed end of the tube through suitable valves (Fig. 275). The invention of this gas pump by H. A. Humphrey in 1909 proved to be of historic interest only although several large installations worked successfully for many years, including one at Potter's Bar, England, and one in South Australia.

Very quickly after the beginning of the century, the steam turbine took first place among the prime movers as the most suitable machine for producing energy in large units both on land and sea. In 1900 the largest steam turbine produced one and a quarter megawatts; by 1923 units of fifty megawatts were under construction and from that time onwards no other type of prime mover was seriously considered for producing large quantities of energy in thermal power stations generating electricity.

The adoption of steam turbines for ship propulsion was more gradual. Unhappily two of the first turbine ships, H.M.S. *Viper* and H.M.S. *Cobra* both met with disaster and were wrecked, the *Viper* on 3 August 1901 and the *Cobra* on 18 September of the same year. In neither case was it established that the turbines were at fault, but understandably marine engineers were for some years afterwards chary of using turbines as the sole means of propulsion. Indeed, some of the early installations of turbines in large liners were limited to the final expansion of the steam exhausted from large reciprocating engines, as in the R.M.S. *Olympic* and R.M.S. *Titanic* (60,000 tons displacement) which were both fitted with steam turbines of 16,000 h.p. each taking the exhaust steam from two 15,000-h.p. reciprocating engines. However, as early as 1904 the Cunard Steamship Company had decided upon steam turbines for the main propulsion of their largest liners, *Lusitania* and *Mauritania*, launched in 1906. Each of these ships was provided with turbines of 70,000 h.p. at a time when the most powerful turbine steamer in service was *The Queen* of only 8,000 h.p., and the *Mauritania* thus was able to win and hold for twenty years the record for the fastest crossing of the Atlantic Ocean.

The progress of steam power in land stations for electricity generation in Britain during the first forty years of the century was reviewed comprehensively by Sir Leonard Pearce in 1939.[32] It can be summed up by comparing the steam engine of 1889 with the steam turbine of 1938 as depicted in Fig. 276, where the outlines of these two machines have been drawn to the same scale. The smaller of the two machines—the turbine alternator—of 1938 produced nearly seven times as much power as the larger reciprocating engine fifty years before. The large steam turbine is inherently more efficient than a small one because the proportion of steam escaping through the clearance between the rotating and stationary parts is reduced. Other economies were made by raising the steam pressure and temperature at the turbine inlet, by reheating—that is, by bleeding off steam between stages, and heating it up again before returning it to the turbine for further expansion—and by superheating the steam, that is, raising its temperature above the saturation temperature corresponding to the boiler pressure. These developments all brought in their wake difficulties in design, construction and maintenance that had to be solved, often at great expense, so that sometimes progress was halted until some difficulty was overcome. Sir Leonard Pearce gave

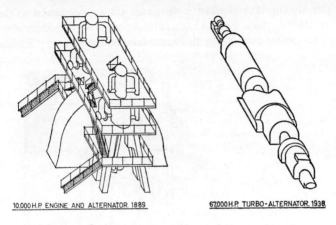

10,000 H.P. ENGINE AND ALTERNATOR. 1889

67,000 H.P. TURBO-ALTERNATOR. 1938.

Fig. 276. Comparison of steam-engine and turbine.

the following figures for the thermal efficiency of steam turbines in the London Power Company's system:

Year	1895	1907	1919	1933	1940
Size of unit kW.	350	6,000	20,000	69,000	100,000
Steam Pressure lb./sq. in.	150	190	200	570	1,350
Steam Temp. ° F.	366	500	600	800	950
Thermal Efficiency of Steam turbines per cent	7·5	18	22	34	38

Similar improvements took place elsewhere and the figures given may be taken as representative of good practice in power station performance at the periods stated.

Among all the promising schemes that were tried during this period without proving generally acceptable was the binary cycle, in which mercury instead of water was used in the high-temperature boiler, the mercury vapour being passed through a turbine before being condensed in a heat exchanger where the latent heat of the condensing mercury was used to raise steam which was expanded through a steam turbine. A large power station built at Hartford in the United States operated successfully on this binary cycle, giving high thermal efficiencies, but the cost and practical difficulties of the system prevented its widespread adoption elsewhere.

415

The development of steam boilers was just as spectacular as that of the steam turbine. For the higher steam pressures and temperatures used for power production the water-tube boiler was most suitable

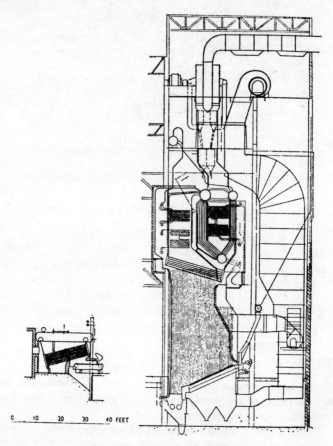

Fig. 277. Steam-boiler plant.

and the growth in size of the boiler with natural circulation, between 1902 and 1940, is illustrated in Fig. 277 where the smaller boiler in 1902 evaporated 12,000 lb./hr. at 150 lb./sq. in. and 366° F., while the larger boiler in 1940 evaporated 550,000 lb./hr. at 1,420 lb./sq. in. and 965° F. The efficiency of the smaller boiler was about 70 per cent and that of the larger about 90 per cent. The improvement was due to the introduction of feed water economizers, air preheaters, the provision of water tubes over the whole surface of the combustion chamber to absorb radiant heat, and accurate controls and instruments.

A new type of steam boiler introduced this century found increasing popularity. This was the forced-circulation boiler of which there

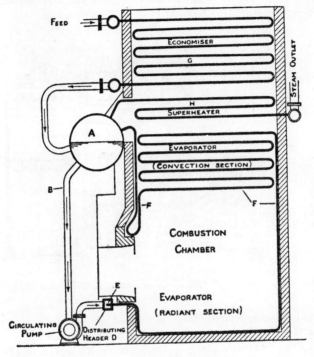

Fig. 278. Lamont steam-generator.

are many different designs. It is claimed for the forced-circulation boiler that smaller bore and hence thinner tubes can be used for high pressures, that the heating surface can be disposed to greater advantage and that fewer drums are required. As against this, pumping equipment is required and absorbs power continuously, and elaborate protection devices must be provided in case of failure of the pumps. Forced-circulation boilers were developed mainly on the Continent of Europe, a typical design being the Lamont boiler illustrated in Fig. 278. Here the pump circulates water round the walls of the combustion chamber via the distributing header and the

O

convection section to the drum A where the steam separates and passes out through the superheater H. The feed water enters the drum after first passing through the economizer G. There were many variations of the forced-circulation principle for steam boilers; for example, in the Loeffler boiler, steam instead of water was circulated by the pump, and after passing through superheater tubes the superheat was used to evaporate water in the boiler drum.

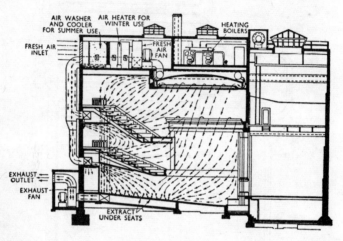

Fig. 279. Ventilation with air-conditioning of a London theatre.

Steam boilers formed only one example where mechanical engineers were concerned with providing machinery and equipment for handling and treating large quantities of air. Air was required not only for boiler furnaces but also for the ventilation of machinery, of mines and of buildings. Equipment had to be designed for filtering air, for heating and cooling it, and for adding or taking away moisture to control its humidity. These operations, termed air conditioning, became the special province of heating and ventilating engineers and some idea of the arrangements involved can be seen from Fig. 279 which relates to the ventilation equipment of a London theatre. In the less temperate zones of the world, the cooling of air required to achieve comfort conditions involved the use of refrigeration equipment. Cooling and conditioning of the ventilating air in offices and public buildings in the United States had become so widespread by 1940 that the total load on the electric supply system in the city of

Washington, D.C., for cooling in summer time was greater than that for heating in the winter.

In the field of refrigeration for most purposes the compression process continued in use, the output from large machines being much increased by improvements in design, particularly of the heat exchangers which form such an important part of any cooling installation. Also in this period, small refrigerating machines began to be developed for domestic use and became so popular that the numbers required justified the adoption of mass-production methods in their manufacture. Many of the compression machines were designed to use a new refrigerant, dichlorodifluormethane—known as Freon[12] —which was non-poisonous and therefore preferable for use in the home.

The most outstanding innovation in refrigeration was the continuous absorption apparatus known as the Electrolux refrigerator. Suitable only for small sizes, it was invented by two Swedish engineers, Platen and Munters, in the first decade of the century. The feature of particular interest to mechanical engineers was that no pump was required and the fluid was kept circulating merely by the application of heat to the generator—'the flame that freezes'. This brought about the extraction of heat at a low temperature from the contents of the cold space to the liquid in the evaporator, and the dispersal of heat at a higher temperature from the absorber and the condenser to the surroundings. The parts of the Electrolux refrigerator are shown in Fig. 280. An important feature was that while the total pressure was practically uniform throughout the system, the partial pressure of the ammonia vapour varied, being high in the condenser and low in the evaporator, the difference being compensated by the presence of hydrogen. Circulation was maintained by convection currents set up by differences in the density of the fluids in different parts of the system.

A new application for large refrigeration plant came into being with the indoor artificial ice-skating rink. Here the floor or stage was covered with some waterproof material in which a grid of steel pipes was laid. Cold brine from the refrigerating plant was circulated through these pipes while they were covered with water, which duly froze, forming an excellent surface for ice skating. This is an example of engineering skill being used to provide the means to a fuller life for many people—without in this case bringing harmful results in its train.

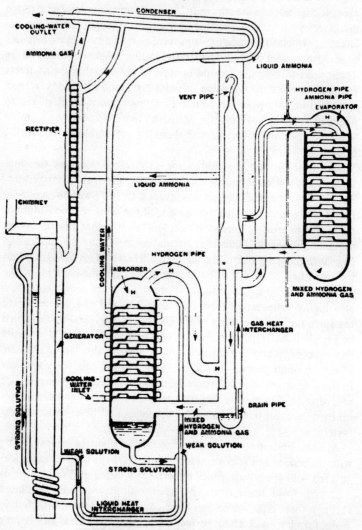

Fig. 280. Electrolux refrigerator circuit.

REVIEW

There can be no doubt that the greatest mechanical innovation of this period was the aeroplane. The achievements of the Wright brothers in the United States in studying the problems of flight and building and flying the first successful aeroplanes have been well told recently by Gibbs-Smith.[2] The enthusiasm, determination and ability applied by the two brothers to solve the problems of mechanical flight in the first five years of this century make a story without parallel in the history of mechanical engineering. It is not generally appreciated that before their first successful aeroplane flight the brothers had taught themselves to become expert glider pilots and that by experimenting with gliders and with models in a wind tunnel they had acquired a knowledge of the scientific aspects and a perception of the practical art of flying that were far ahead of their time. By contrast, the very slow evolution of the practical powered aeroplane in Europe has been described as the outstanding mystery in the history of aviation. Compare the achievement of the Wright brothers, who in the year 1905 made 49 flights, several of more than half an hour's duration, with that of the European aeronauts, none of whom was able to stay in the air for a single minute until November of 1907.[*]

Gibbs-Smith considers that there was no conceivable reason why an aeroplane should not have been flown in Europe by 1905 or at the latest by 1906, and had this been done the resulting developments might have transformed the First World War, and thereby changed the history of the world. Nevertheless, during and after the war, aeroplane development proceeded apace in Europe, though it was in the United States that the all-metal aircraft first came into general use and it was there that the first large four-engined airliners such as the Boeing 307 Stratoliner in 1938 (Fig. 281) were designed and used.

The success of the aeroplane is in marked contrast to the development of the airship, which achieved considerable success until a number of disasters led to the abandonment of this form of aerial transport before 1940.

Towards the end of this period experimenters in several countries were engaged in developing aircraft with rotating wings, both the type in which the wings are rotated by power—now called the helicopter—and the type where the wings are rotated only by the forward motion produced by the traction propeller—now called the autogyro.

[*] Loc. cit., p. 321.

Neither of these forms of aircraft came into general use before 1940 though they have since become well established. They are particularly suited for short journeys and for landing and taking off in a confined space.

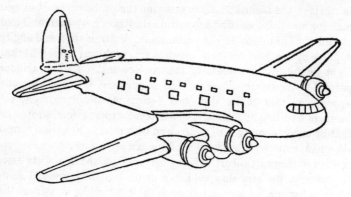

Fig. 281. Boeing airliner, 1938.

Developments in aeronautics have been closely associated with fundamental studies of fluid mechanics based upon the concept of a boundary layer of the fluid existing wherever it is adjacent to the surface of a solid. Such studies were started by Prandtl in Germany, but they spread quickly to other countries and it was soon found that their results could be applied usefully to other fields besides aeronautics. For example, the whole subject of heat transfer by convection, as in boilers, furnaces, condensers and feed water heaters, was transformed once the boundary layer concept was applied to it, and in consequence the estimation of the conditions required for heat transfer became a science where before it had been an empirical art.

All such developments, however, were not successful and one proved to be a costly failure. This was the curious way of applying the boundary layer theory to ship propulsion which was proposed in 1924 by A. Flettner in Germany for propelling a ship by means of one or more rotating vertical cylinders mounted on the deck. These rotating cylinders took the place of the sails on a sailing ship and by their rotation made use of the so-called 'Magnus effect' resulting in a layer of air being carried round the cylinders, so that the pressure difference across them was much greater than it would have been if the cylinders had not been rotated. It was thought that the idea would have application for tramp ships and fishing boats, but though two ships were built to operate in this way, it was found more costly than alternative means of propulsion.

Next to the aeroplane, the machine which underwent the greatest development and made the most impact upon civilization was the automobile. The machine that will go down in history as typical of

Fig. 282. Ford Model T automobile.

the period is the Model T Ford (Fig. 282) of which 15 million were built between 1908 and 1927. When first produced this model was so simple and so much in advance of its time that it seemed that its popularity would last for ever; it could be started without a battery with its flywheel magneto; it had a two-speed epicyclic gear box, a side valve engine with a single cylinder block for the four cylinders, a detachable cylinder head, and a vanadium steel crankshaft. It was to make this machine that Henry Ford[33] built his vast factory at Dearborn that produced automobiles first in hundreds, then in thousands and finally in millions. Methods of assembly on a conveyer belt were developed for putting it together; plate glass for the windows was made by a continuous process, and above all methods of high-speed machining, forging, casting and forming metals were developed which from that time onwards provided a stimulus for improvements that shows no sign of slackening to this day. The developments were not confined to Ford's plant or even to America, though for many years they were more in evidence in the United States than elsewhere. An example of the trend towards continuous manufacture in automatic plant was the A. O. Smith factory at Milwaukee, Wisconsin, where automobile chassis frames were made, assembled, cold riveted and stoved on a conveyer at the rate of one every six seconds.

Another significant development of this era was the Keller Milling machine devised for machining the large steel dies used for pressing steel bodies, doors and wings for automobiles. These machines were able to reproduce exactly in steel the shape of a hand-made model —usually of wood—by traversing a sensing probe over the surface of the model, thereby actuating either electrically or hydraulically the controls of the milling head which cut out the metal to the same shape as the model.

The automobile and the aeroplane depended for their success on suitable internal combustion engines to drive them. For both these applications the high-speed piston engine, using petrol as the fuel, was the most popular. Improvements were made rapidly in the quality of the fuel and in the design of these engines so that they became more efficient (producing more energy per unit of fuel consumed) and of greater specific power (producing more energy per unit of weight of engine or of space it occupied). For some purposes, such as for driving lorries, buses and tractors, the petrol engine, after a period of popularity—up to about 1930—gave place to the diesel oil engine, which was more efficient, though more costly, heavier, noisier and less flexible than the petrol engine.

The diesel engine also replaced the gas engine where it had not already been replaced by the electric motor, which became the most popular source of power where an electricity supply was available. Gas engines fell rapidly out of favour, whether supplied by town's gas from the mains, by gas producers, or by waste gas from blast furnaces, their places being taken by electric motors, diesel engines or—for the largest power units—steam turbines. The internal combustion engine was applied to many other uses besides transport where compactness and mobility were essential requirements for the power unit. Mobile cranes, stackers and fork-lift trucks were developed for use in factories and warehouses, a feature of these machines being that the same power unit was used both for moving the whole machine and then, by disengaging the engine from the transmission and engaging another gear, for lifting and lowering the load. This idea of a 'power take-off' from a vehicle, first used on the steam traction engine of the previous century, was the basis of the popularity of the tractor for agricultural purposes, and when applied to tractors powered with internal combustion engines it led to the development of a large variety of machines of which the combine harvester and the potato digger were examples.

One of the most difficult problems that had to be solved was how to propel a tractor over uneven ground that might be hard and rocky or of soft sandy soil or even of mud. For the lighter, faster tractors large pneumatic tyres with specially formed treads were frequently the best solution, but for the heavier tractors the crawler type of caterpillar vehicle was developed, laying its own track and spreading its weight over a much greater surface than was possible with wheels. This method of propelling a vehicle was used by the 'tanks' or armoured fighting vehicles of the First World War and was developed to a fine degree between the wars, particularly in Germany, Russia and the United States. In the United States, and in Germany, the tractor was used as the basis for developing a variety of powered machines for civil construction work such as road-building. Earth-moving machines such as bulldozers, dumpers, scrapers and excavators, that are now a common sight where civil engineering work is in progress, supplanted the manual methods of road-building that had been hallowed for centuries. The need to employ large gangs of navvies to dig with spades had gone for ever, and the extent to which powered machinery was now used to alter the landscape has since then become one of the mechanical marvels of our time.

Of all the developments in mechanical engineering that took place in this period, a much larger proportion than ever before resulted from applying scientific principles to the design and manufacture of machinery, as seen in the evolution of aircraft and the mass production of automobiles, in the improvements made to boilers, turbines, and internal combustion engines, and in the development of welding and in the use made of scientists in the workshop. Often the scientists or scientifically-trained engineers were brought in to explain and remedy some unexpected failure, a task at which they succeeded often enough for it to be realized that if they had been consulted before the machinery had been built, the failures might never have occurred. Nevertheless, engineers learn a great deal from their failures, though it is better to induce these to happen in the laboratory rather than for them to occur unexpectedly on full-scale machinery. Several examples of how engineers learn from failures were given by L. W. Schuster[34] from whose paper Fig. 283 has been

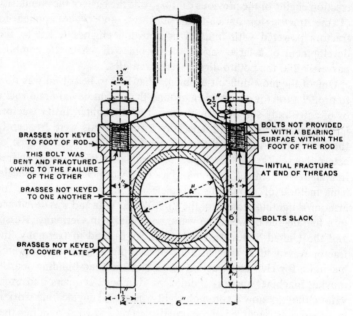

Fig. 283. Example of faulty design. Connecting rod bearing.

chosen. By publishing information of this sort to professional engineers the Institutions added to the total stock of knowledge and professional skill, no less than when they notified their members of new designs or inventions.

REFERENCES

(for abbreviations see List of Acknowledgments)

1. Brooks, P. W., 'The Development of the Aeroplane', *Journal of the Royal Society of Arts*. Cantor Lecture, January 1959.
2. Gibbs-Smith, C. H., *The Aeroplane*. Sc. Mus. Pubn., H.M. Stationery Office, London, 1960.

3. Stodola, A., *Steam and Gas Turbines*, trans. by L. C. Loewenstein, 2 vols. McGraw Hill, New York, 1927.

4. Andrade, E. N. da C., 'On the viscous flow in metals', *Proc. Royal Society*, 'A' vol., lxxxiv, 1910.

5. Bailey, R. W., 'Utilisation of creep test data in engineering design', *Proc. I.M.E.*, vol. 131, 1935, p. 131.

6. Smith, F. E., 'Plant for the production of petrol by the hydrogenation of bituminous coal', *Proc. I.M.E.*, vol. 133, 1936, p. 139.

7. Barber, A. T., and Taylor, A. H., 'H.P. plant for experimental hydrogenation processes', *Proc. I.M.E.*, vol. 128, 1934, p. 5.

8. Haigh, B. P., *Journal of the Institute of Metals*, vol. 18, 1917.

9. Coker, E. G., 'Elasticity and Plasticity', *Proc. I.M.E.*, 1926, p. 897.

10. Rolt, F. H., *Gauges and Fine Measurements*, 2 vols. Macmillan, London, 1929.

11. Strassman, W. P., *Risk and Technological Innovation*. Cornell U.P., 1959.

12. Taylor, F. W., *On the Art of Cutting Metals*. American Society of Mechanical Engineers, 1907.

13. Schlesinger, G., 'Machine Tool Tests and Alignments', *Proc. I.M.E.*, vol. 138, 1938, p. 59.

14. Woodbury, R. S., *History of the Grinding Machine*. M.I.T.P., 1959.

15. Sperry, E., *The Gyroscope through the ages*. Sperry Gyroscope Co., U.S.A.

16. Inglis, C. E., 'Gyroscopic principles and applications', *Proc. I.M.E.*, vol. 151, 1944, p. 223.

17. Hodgson, J. L., and Robinson, L. L., 'Development of automatic combustion control systems for industrial and power station boilers', *Proc. I.M.E.*, vol. 126, 1934, p. 59.

18. Parsons, R. H., *Development of the Parsons Steam Turbine*, p. 142. Constable, London, 1936.

19. Nyquist, H., 'The Regeneration Theory', *Bell System Technical Journal*, 11, 1932, p. 26.

20. Callendar, A., Hartree, D. R., and Porter, A., 'Time lag in a control system', *Phil. Trans.*, A 235, 415, 44, 1936.

21. Bauersfeld, W., 'Projection planetarium and shell construction', *Proc. I.M.E.*, vol. 171, 1957, p. 75.

22. Goldstein, S., (ed.), *Modern Developments in Fluid Mechanics*. O.U.P., 1938.

23. Davy, M. J. B., *Aeronautics: Lighter-than-air aircraft*. Sc. Mus. Pubn., H.M. Stationery Office, 1950.

24. Barwell, F. T., *Lubrication of Bearings*. Butterworth's Sc. Pubns., London, 1956.

25. Sinclair, H., 'Problems in the transmission of power by fluid couplings', *Proc. I.M.E.*, vol. 139, 1938, p. 83.

26. Constantinesco, G., *Theory of Wave Transmission*. Haddon, London, 1922.

27. Booth, H. C., 'The origin of the vacuum cleaner', *T.N.S.*, vol. xv, 1934–5, p. 85.

28. Smith-Clarke, G. T., 'Mechanical Breathing Machines', *Proc. I.M.E.*, vol. 171, 1957, p. 52.

29. Lanchester, F. W., 'The gas engine and after', *Proc. I.M.E.*, vol. 136, 1937, p. 195.

30. Ricardo, Sir H. R., *The High-Speed Internal Combustion Engine*, 4th edn. Blackie, London, 1953.

31. Purdie, W. H., 'Thirty years' development of opposed-piston propelling machinery', *Proc. I.M.E.*, vol. 162, 1950, p. 446.

32. Pearce, Sir S. L., 'Forty years' development in mechanical engineering plant for power stations', *Proc. I.M.E.*, vol. 142, 1939, p. 305.

33. Ford, H., *My Life and Work*. Garden City Publishing Co., 1926.

34. Schuster, L., 'The investigation of the mechanical breakdown of prime movers and boiler plant', *Proc. I.M.E.*, vol. 124, 1933, p. 333.

BIBLIOGRAPHY

Barwell, F. T., *Lubrication of bearings*. Butterworth's Sc. Pubns., 1956.

Bevan, T., *Theory of Machines*. Longmans Green, London, 1939.

Brame, J. S., and King, R. O., *Fuel*, 4th edn. Arnold and Co., London, 1935.

Carpenter, H., and Robertson, D., *Metals*, 2 vols. O.U.P., 1939.

Crowther, J. G., *Discoveries and Inventions of the Twentieth Century*, Routledge and Kegan Paul, London, 1955.

Dalby, W. E., *Steam Power*, 2nd edn. Arnold and Co., London, 1920.

Davy, M. J. B., *Aeronautics: Heavier-than-air aircraft*, Part I. Historical Survey, Sc. Mus. Pubn., H.M. Stationery Office, London, 1949.

Davy, M. J. B., *Aeronautics: Lighter-than-air aircraft*. Sc. Mus. Pubn., H.M. Stationery Office, London, 1950.

Ewing, J. A., *The Steam Engine and other Heat Engines*, 3rd edn. C.U.P., 1920.

Ewing, J. A., *The Mechanical Production of Cold*. C.U.P., 1908.

Field Foster, P., *The Mechanical Testing of Metals and Alloys*. Pitman, London, 1948.

Fishenden, M., and Saunders, O. A., *Calculation of Heat Transmission*. H.M. Stationery Office, 1932.

Freedman, P., *Principles of Scientific Research*. Macdonald, London, 1949.

Gibson, A. H., *Hydraulics and its applications*, 4th edn. Constable, London, 1930.

Gordon, Lord Dudley, 'Recent Developments in Refrigeration', *Proc. I.M.E.*, vol. 149, 1943, p. 49.

Gibbs-Smith, C. H., *The Aeroplane*. Sc. Mus. Pubn., H.M. Stationery Office, London, 1960.

Goldstein, S. (ed.), *Modern Developments in Fluid Mechanics*. O.U.P., 1938.

Goodman, J., *Mechanics Applied to Engineering*. Longmans Green, London, 1930.

Hoover, T. J., and Fish, J. C. C., *The Engineering Profession*. Stanford U.P., 1941.

Den Hartog, J. P., *Mechanical Vibrations*. McGraw Hill, New York, 1940.

Karman, T. von, and Biot, M. A., *Mathematical Methods in Engineering*. McGraw Hill, New York, 1940.

Kelly, T., *George Birkbeck*. Liverpool U.P., 1957.

Lanchester, F. W., *Aerodynamics*. Constable, London, 1907.

Lanchester, F. W., *Aerodonetics*. Constable, London, 1908.

Lea, F. C., *Hydraulics*, 6th edn. Arnold and Co., London, 1938.

McAdams, W. H., *Heat Transmission*. McGraw Hill, New York, 1933.

Midgely, T., and Boyd, T. A., 'Chemical Control of Gaseous Detonation', *Journal of Society of Automotive Engineers*, 1922.

Morley, A., *Strength of Materials*, 9th edn. Longmans Green, London, 1940.

Moyer, J. A., and Fittz, R. U., *Refrigeration*. McGraw Hill, New York, 1932.

Newitt, D. M., *Design of H.P. Plant and properties of fluids at high pressure*. O.U.P., 1940.

Ower, E., *Measurement of Air Flow*, 2nd edn. Chapman and Hall, London, 1933.

Parsons, R. H., *A History of the Institution of Mechanical Engineers 1847–1947*, I.M.E., London, 1947.

Piercy, N. A. V., *Aerodynamics*. English U.P., London, 1937.

Porter, A., *An introduction to Servo-mechanisms*. Methuen, London, 1950.

Prandtl, L., and Tietjens, O. G., *Applied hydro- and aero-mechanics*. McGraw Hill, London, 1934.

Prandtl, L., *The Physics of Solids and Fluids*, 2 vols., 2nd edn. Blackie, London, 1936.

Pye, D. R., *The Internal Combustion Engine*, 2 vols., 2nd edn. O.U.P., 1937.

Ricardo, Sir H. R., *The H.S. Internal Combustion Engine*, 4th edn. Blackie, London, 1953.

Rollason, E. C., *Metallurgy for Engineers*. Arnold and Co., London, 1939.

Rolt, F. H., *Gauges and Fine Measurements*, 2 vols. Macmillan, London, 1929.

Southwell, R. V., *Theory of Elasticity*. C.U.P., 1936.

Stodola, A., *Steam and Gas Turbines*, trans. by L. C. Loewenstein, 2 vols. McGraw Hill, London, 1927.

Tool Engineers' Handbook (American Society of Tool Engineers). McGraw Hill, London, 1951.

Walshaw, A. C., *Heat Engines*, 4th edn. Longmans Green, London, 1956.

CHAPTER IX

Dawn of the Nuclear Age and Space Travel
1940–1960

T he last twenty years are still so near to us that it is difficult to appraise the technical achievements of that time with a proper historical perspective. All we can do is to point out that some of the recent inventions and developments seem likely to deserve a place in the history of this period when it comes to be written a hundred years hence.

Once again it seems that it is in the field of fluid machines that developments have been most remarkable, so we will consider them first.

Most spectacular mechanical achievements were made by the Russians who first succeeded in putting a series of man-made satellites into orbit round the earth. The mechanical problems encountered during the years of research that preceded these events have not yet been disclosed to the Western world, though some have been successfully and independently overcome by workers in the United States, who followed the Russian achievement by putting some smaller satellites into orbit. The British contribution in this field was the construction of a gigantic radio telescope at Jodrell Bank, so far the only instrument in the world capable of tracking these space flights. When it is realized that less than thirty years ago responsible aeronautical engineers were asserting that flight at speeds above that of sound (700 miles/hour) was impossible, and that to escape from the earth's atmosphere a satellite would have to start its flight with a velocity more than fifty times greater (7 miles/sec.) we get some idea of the tremendous progress that has been made in this branch of engineering.

Aircraft speeds exceeding the speed of sound are now commonplace, and became so soon after the introduction of the jet propulsion engine during the Second World War. Fig. 284 shows diagrammatically one of these engines of the type used by Whittle in his pioneer work on the jet propulsion of aircraft. In this the air was drawn through a centrifugal compressor, delivered to the combustion chamber and then expanded through the turbine where sufficient power was produced to drive the compressor, and thus the aircraft was propelled entirely by the reaction produced from the exhaust gases issuing through the jet pipe at the rear. Similar machines with axial-flow compressors, such as the de Havilland Comet airliner

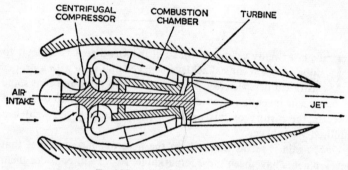

Fig. 284. Jet propulsion engine.

which went into service in 1952, have been equally successful. When such jet propelled aircraft operated at supersonic speeds (i.e. above the speed of sound) the compressor and turbine became redundant, since the forward speed of the aircraft was then so high that the air was rammed into the inlet tube sufficiently fast to build up a pressure to maintain continuous combustion. Engines of this type, known as ram-jet motors, could only be used as the main propulsion units on aircraft where other means were provided for bringing the craft up to operating speed. They were also used on the tips of helicopter rotors to augment their speed.

A similar principle was used in the engines of the German flying bombs in the Second World War; they used a pulse-jet engine in which the ramming effect was periodically interrupted by hinged shutters closed intermittently by explosions in the combustion tube. Pulse-jet engines have also been used on the blade tips of helicopter rotors. In many of the slower modern aircraft the gas turbine was used to drive a propeller, the thrust of which was augmented by the jet from the exhaust. Such machines, known as turbo-prop or prop-jet aircraft have been very successful during the last ten years—the Vickers Viscount being a typical commercial airliner of this type.

At the end of the Second World War it was generally expected that the gas turbine would quickly supersede the steam turbine for power stations, for industrial purposes, for ship propulsion, for locomotives and even for automobiles, as it rapidly did for aircraft, but even after the most intense research work over a period of twenty years, this has not come about, largely because of the (so far) insurmountable difficulty of operating continuously the high-speed machinery at the very high temperatures required for the economic operation of

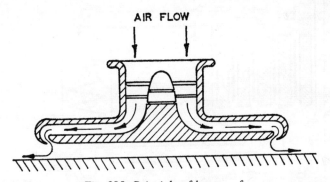

Fig. 285. Principle of hovercraft.

gas turbines. Many large units of different kinds and for different purposes were built and operated with disappointing results, so that the steam turbine, the diesel engine and the petrol engine have remained the principal heat engines.

The solution of the gas turbine problem required materials of qualities superior to anything that had yet been produced and a great deal of research activity among metallurgists had been going on to this end. One of the most significant contributions made by mechanical engineers was in the field of heat transfer, particularly of fluids moving in different ways past metal surfaces. These studies have ranged from the behaviour of liquid metals, water, oil, and air under streamline flow, up to the study of shock waves under conditions of supersonic flow and pulsating flow.

The greater awareness of the phenomena of fluid flow was responsible for some interesting mechanical devices such as the Hovercraft aeroplane (Fig. 285) which could move about, a little above ground level, supported by a cushion of air, and similar machines for road and rail that worked on the same principle. Other results which followed were the more general use of air-lubricated bearings for machinery, and the application of the vortex flow theory to the design of turbine blading. The latter was designed with a twist along its length, so that there was no radial pressure gradient caused by the swirl of the gas. This reduced losses by reducing the tendency for radial flow.

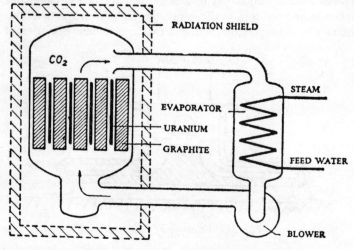

Fig. 286. Diagram of Calder Hall nuclear power-station.

Another great achievement in engineering during this period was the construction of nuclear reactors and their use for generating electricity. At first reactors were built as bombs for warlike purposes, to produce nuclear explosions, and later for controlled nuclear reaction, the heat released being used to produce steam to operate turbo-alternators for generating electricity. Reactors were also used to produce isotopes that had many applications in both industry and medicine. Atomic fission on a prototype scale was first carried out by Fermi, in Chicago, in 1942, using uranium and uranium oxide as the fuel element and graphite as a moderator. By this means he was able to demonstrate experimentally what had already been predicted on theoretical grounds, namely, that by atomic fission a self-sustaining and controllable reaction could be initiated. Subsequently, atomic piles, as these reactors were called, were built in Great Britain, the United States, and elsewhere and by 1957 the first nuclear power station was completed at Calder Hall, England (Fig. 286). By 1960 nuclear power stations were going up in Britain, the United States, Russia, Italy, France and Japan, and a nuclear propelled submarine had been operated for some years by the United States Navy.

Great mechanical skill was displayed not only in the design of the reactors but also in the factories where the uranium fuel was processed and prepared. The radioactive material being extremely dangerous to life, means had to be contrived for constructing and operating the equipment by remote control, for which special mechanical manipulators were devised. The hazards of nuclear radiation were minimized by surrounding the equipment with massive walls of concrete, by extensive use of automatic control devices, and by transferring the heat in the reactor to an exchange medium such as carbon dioxide or water, having a low neutron absorption, a high specific heat and great thermal stability (Fig. 287).

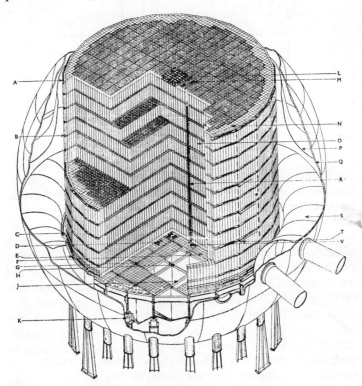

Fig. 287. Section through reactor core.

The steam turbine machinery used for generating electricity in nuclear power stations was conventional; in fact it was less advanced technically than the corresponding equipment in power stations using fossil fuel, where the tendency to use still higher steam pressures and temperatures, and larger units, was continuing. Single units of 50 MW., 120 MW. and 250 MW. were not uncommon, and by 1960 designs were being prepared in England for a single generating unit of 550 MW.

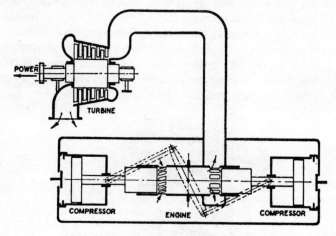

Fig. 288. Free-piston hot-gas generator.

An innovation that came to be associated with the gas turbine was the free piston diesel engine (Fig. 288) in which there were no cranks or rotating parts, the energy released by the explosions being used directly to compress air in cylinders in line with the working cylinder. Once started, the pistons continued to fly back and forth so long as fuel was supplied to the injectors. Developed originally by Pescara in France for compressing air for construction work and other purposes, it was found to have a special value as a gas generator for a gas turbine, since a higher efficiency of combustion and compression could be obtained with this reciprocating unit than with any rotary machine so far produced. A further advantage was that with such an arrangement extremely high ratios of compression could be used, 30/1 or more, resulting in very high temperatures of combustion so that it was claimed that cheaper fuels than would have been possible with alternative equipment could be burnt successfully.

Another innovation was the Philips refrigerating machine for making liquid air at atmospheric pressure. It consisted of an enclosed reciprocating compressor with a co-axial displacer (*see* Fig. 289) so arranged that the enclosed working fluid—hydrogen—passed in

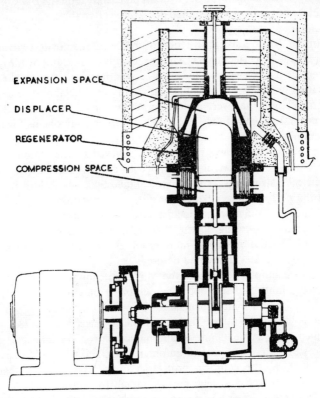

Fig. 289. Philips refrigerating machine.

turn through a compression space, cooler, regenerator and expansion space and back again, thus following the events of the Stirling hot-air engine cycle in reverse, and thereby producing intense cold in the expansion space, the outer walls of which were used for cooling atmospheric air down to the temperature at which it liquefied. This machine was developed by the Philips Co. in Holland as the result of researches originally directed towards improving the Stirling cycle hot-air engine, an endeavour that has not yet achieved commercial success. The liquid-air machine however proved particularly useful in providing hospitals and laboratories with an independent supply of uncontaminated liquid air, the only services required being supplies of cooling water and electricity. After an interval of about ten minutes from switching on, the machine provided a continuous supply of liquid air through the drain pipe (shown on the right of the figure) until at length it required de-frosting after about sixteen hours' running.

In the design and construction of machines the greatest advances were made by perfecting mechanisms already known and by the more extensive use of electrical aids and accessories; for example, the calculating machine was developed into a number of elaborate computing machines, that were almost entirely composed of electrical elements and depended very largely on electronic components for their operation. Mechanical engineers came to use such computers for design calculations and were thereby relieved of the tedious work of computation. They also depended to an increasing extent upon electrical instruments for experimental investigations and for recording the performances of their machines in operation. As machines became more and more automatic, it was usually electrical controllers and electronic sensing devices that were chosen to replace manual control, because they had a more rapid response to change than any alternative devices that could be used.

Many examples could be given to illustrate the way that electrical elements infiltrated into machine design so that the essential part on which the machine depended for its operation or control was frequently electrical rather than mechanical. Electrical control was applied to instruments and to automatic controls for machines and processes, to computers and automatic machine tools and copying machines, to the cathodic protection of metal structures against corrosion, to the magnetic particle clutch for automobiles, to the electro-magnetic mercury pump used in nuclear engineering and to the electronic control of machine tools by means of punched cards or magnetic tape, which resulted in the 'digital machining' that introduced new standards of accuracy and speed in the cutting and shaping of metals.

Digital machining was an innovation that involved a real 'breakthrough' in manufacturing methods for it meant that no machine drawings of the conventional kind were required for the machining of a component. Instead, a set of figures was worked out showing the co-ordinates of each point on the surface of the component in each of three dimensions. Adjacent points were separated by 'digits' representing the accuracy required. The figures supplied to the machine represented 'instructions' which controlled the movements of the tools and slides of the machine tool so that when they had been followed the finished component had the exact dimensions prescribed by the 'instructions'.

Most of the processes of machine manufacture could now be carried out by machines that did not depend upon the manual skill of the operator for the speed or accuracy of the work done. Such automatic machines were usually very costly and production in very large quantities had to be assured to justify their cost, so that they were used for the making and assembly of such articles as automobiles, electric motors, vacuum cleaners and refrigerators where tens or hundreds of thousands of exactly the same product were required.

The same techniques of production by automatic machining were extended rapidly to other products with the result that the skilled engineering craftsman tended to be required only for making 'the machines that make machines', for making machines in small quantities, and for making repairs and modifications to machines of all kinds. One result of these changes in the future may be that skilled men will no longer be required to do monotonous repetition work, for as soon as a task becomes onerous, it will be possible to make a machine to do it automatically.

An example of this trend was to be found in the portable chain-tooth saw (a most ingenious innovation entailing a complete departure from methods that had been used since the Stone Age) which was fast replacing the traditional woodman's axe and the two-man hand-saw for felling trees. The chain saw consisted of a thin metal frame furnished with a steel roller chain, carrying saw teeth attached at intervals along its length, the saw teeth being slightly wider than the chain or the frame to ensure that the saw cut was sufficiently wide to allow the frame to enter it. It was driven through one of the chain sprockets by a very small two-stroke petrol engine mounted on the frame. This speeded up both the task of tree felling and the cutting up of logs and also removed the gruelling labour and monotony that has been associated with this work since time immemorial.

In the shaping of metals, a preference for using forming rather than cutting processes developed. Cutting processes produce waste material and are inherently slower than forming processes such as pressing, rolling and hammering. Whereas in the past forming methods had to be carried out on metals at a red heat, because sufficient power was not available for working the metal cold, now a variety of methods were being developed for cold forming, cold rolling, cold forging, and cold pressing that could produce products, with great accuracy, and with no wastage, in a fraction of the time required with other methods. Also the metal in the finished component was usually in a better metallurgical state as a result of the cold forming operations. Some cold forming was done with the sudden impact of under-water explosions, other processes were purely mechanical using either steady pressure or repeated hammering or both together. The manufacture of finned metal tubing, in which the fins were integral with the wall of the tube was but one example of a product produced by these methods. Tubing with fins that are longitudinal or spiral could now be made by extrusion through a die, while some steel tubing with radial fins was made by a thread rolling process. Finned tubing was now used in vast quantities for air-conditioning plants in aircraft, railway trains and buildings and for heat exchangers in industry, particularly in nuclear engineering applications.

The advent of nuclear engineering brought renewed attention to the importance of considering hazards to health in engineering work quite apart from those arising from radiation. The acceptance by engineers of the dangers to health from hazards such as dust, noxious substances, noise, vibration and sudden changes of temperature or pressure were challenged by medical practitioners, some of whom became more interested in maintaining the health of patients than curing diseases. The study of the health of men at work became their main preoccupation; they instituted periodic medical examinations of work people engaged in hazardous tasks and, further, began experimental researches to discover the physical and psychological causes of illness, fatigue, boredom and lack of interest at work. Such studies have led to changes in the location and display of instruments, the position of control levers and seating on machines, and to the general use of goggles, helmets, gloves and protective clothing wherever they were considered necessary.

A recent innovation that reduced the danger to health from airborne dust during the fettling of iron castings was the introduction of the hydro-blast in place of sand blasting with compressed air. In the new process the abrasive sand or steel shot was propelled by high-pressure, high-velocity water instead of air, thus reducing to negligible proportions the danger of silicosis that had always been associated with fettling.

Great changes had been taking place in steelmaking, the most remarkable being the general adoption of the oxygen blowing of steel in the furnace. This was a partial return to the Bessemer process in which air was blown through the molten metal in a furnace, but the new process had the added advantage that no impurities or nitrogen were brought in contact with the steel. It only became practical when cheap 'tonnage oxygen' came on the market as a result of improvements in chemical engineering techniques which made it possible to supply liquid oxygen cheaply by the ton (tonnage oxygen).

It is now claimed (1961) by steel masters in Britain and the U.S.A. that some success has been achieved in the continuous casting of steel. It is a process in which a stream of molten steel pours continuously from the base of a furnace into a water-cooled agitated mould from the base of which it emerges as a white hot bar and is rolled immediately into plate, strip, or whatever section is required. The process can be quite continuous and avoids the necessity of reheating ingots and billets that has been customary since ancient times.

The use of plastic materials in engineering has increased greatly in the last twenty years. After the war it was anticipated that they would replace metals for applications where the light weight and resistance to corrosion of plastics would be of special value. It was envisaged that the fuselages and wings of aircraft, motor-car bodies and ships' hulls would be made of plastics, but so far this has occurred only to a limited extent—the bodies of some automobiles and caravans and of lifeboats and dinghies have been made of them, usually reinforced with glass fibre. Some unexpected applications of plastic materials have been made that may be significant in the future for mechanical engineering; for instance, the use of plastics for making flexible fuel tanks for aircraft, for road transport and for towing underwater behind a ship, have all been developments of an unusual kind. In aircraft, flexible fuel tanks in the wings reduced the fire and explosion hazards in the event of forced landing; for road or rail transport flexible tanks could be carried when full on flat platform trucks or bogies and when discharged they could be rolled up into a parcel to occupy only a small space while being returned empty. Something similar was the 'Dracone' barge (Fig. 290) devised by Professor W. R. Hawthorne of Cambridge, England in 1956. This was a very large 'sausage' or 'sock' made of nylon proofed with rubber and towed underwater when full of oil and rolled up into a parcel that was carried on the ship's deck for the return journey. The studies made by Hawthorne and his collaborators of the stability of these vessels under tow introduced some new and unusual features for the mechanical engineering designer.

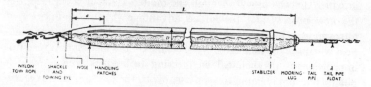

Fig. 290. Dracone flexible barge.

That plastic materials were used also for making ropes, piping, cams, gears and bushes calls for no special comment since it could have been foreseen they would be likely substitutes for hemp or metal as soon as materials were produced that had superior physical properties, such as higher tensile strength, more ductility and better resistance to corrosion. What was more unexpected was the discovery that certain plastic materials such as PTFE—polytetrafluoroethylene—have coefficients of friction that are even lower (0·05) than that of a well-lubricated journal bearing. Another virtue of certain plastics is their transparency, which makes them suitable for making lenses and for replacing glass with a material that can be easily machined, shaped and welded. In this way such plastic materials are particularly useful for experimental work where it is valuable to be able to see what is happening inside a machine or model.

A new tool for engineering investigation was provided by certain plastic materials such as an ethoxyline resin which possessed not only photo-elastic properties (i.e. of displaying stress patterns by colour bands when polarized light was passed through it) but was also able to retain at room temperature locked up stresses previously induced in the material by subjecting it to loading at a suitable higher temperature and then cooling under load. These two properties made it possible to study the stress system that had been imposed on a three-dimensional specimen by the so-called 'frozen stress technique' in which the three-dimensional specimen, after loading and cooling, was cut into slices, each slice having its stress pattern examined under polarized light.

Another unforeseen development in experimental stress analysis that depended upon plastic material was the use of a special plastic paint on a metal surface, which under load caused cracks to appear in the paint, the width and disposition of the cracks giving a measure of the stress in the metal surface beneath.

The most elegant method of stress analysis which came into general use in this period was by the use of electrical resistance strain gauges. Various types of extensometer or strain gauge had been used before to measure the strain at selected parts of machines or structures under load, but the idea of cementing small pieces of resistance wire on to the surface of members under load, developed in the U.S.A. during or just before the Second World War, has now resulted in one of the most important changes in the practice of mechanical engineering. By this means the working stress in any part of a design can be

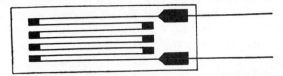

Fig. 291. Electric-wire resistance strain-gauge.

checked by gluing a gauge no larger than a postage stamp to the surface of the metal, in the region where it is desired to know the working stress while the part is under load. It had been shown by Lord Kelvin in 1856 that metal wires carrying an electric current suffer a change of electrical resistance when subjected to a tensile load and that the change of electrical resistance was directly proportional to the tensile strain in the wire, but it was not until this period that the phenomenon was used to obtain direct measurements of the strain of parts of machines and structures while under operational or test conditions of loading. The method used was to cement a grid of resistance wire (Fig. 291) between two thin pieces of paper, the size of a small postage stamp, and then to fix one side with an adhesive bond directly to the metal surface where the strain was to be measured. The gauge could then be connected into an electrical circuit where the change of electrical resistance gave a quantitative measure of the strain. This method could be used for the rapid weighing of an aircraft on landing in an airfield or for the detailed experimental analysis of the stresses, both static and dynamic, in all parts of a machine under load, and for many other purposes.

All the innovations that have been mentioned in this chapter have had a scientific basis and have required considerable technical knowledge and skill for their development. Progress in mechanical engineering has come to depend much more than heretofore on an appreciation of the significance of scientific discoveries. Accordingly it is not surprising that more people are being educated and trained in the methods of science and technology in the hope that they may make fuller use of what is already known. In Britain and elsewhere there has been since 1940 an enormous increase in the facilities for technical education and a larger output of trained engineers. More universities have been established in Britain and a number of the larger technical colleges have been up-graded to the status of Colleges of Advanced Technology where students are enrolled only for higher courses—of at least the standard of a new award—the Diploma in Technology—that is equivalent to the Diplomas that have been awarded for many years to professional engineers on the Continent. While the numbers of trained engineers both in Universities and Technical Colleges have greatly increased, the engineering departments of British Universities and Colleges of Advanced Technology have recognized the need for changes in curricula and shown their awareness of the need for greater specialization in mechanical engineering by establishing courses and professorships in nuclear engineering, production engineering, fluid mechanics and applied mechanics. This last, which has been the most general, is probably the most significant. It recognizes not only the importance of vibration analysis and noise control but more importantly, the rise of control engineering—now the concern of every kind of professional engineer. Control engineers of the present day have to face the solution of intractable non-linear problems. The *raison d'être* for the concern of the mechanical engineer with control problems, is that whatever the nature of the units comprising the control system, be they electrical, pneumatic, hydraulic or mechanical, the unifying theme is dynamics, and it seems that this is also the central theme of some of the most important recent developments in mechanical engineering.

BIBLIOGRAPHY

(for abbreviations see List of Acknowledgments)

Aughtie, F., 'Electrical resistance wire strain gauges', *Proc. I.M.E.*, vol. 152, 1945, p. 213.

Baxter, A. D., 'Rockets in Space', *Journal of Stephenson Engineering Society*. King's College, Newcastle-upon-Tyne, vol. 2, no. 4, 1959–60, p. 7.

Bowden, F. P., 'Recent studies of metallic friction', *Proc. I.M.E.*, vol. 169, 1955, p. 7.

Bristow, J. R. (et alii), 'Use of wire resistance strain gauges in automobile engineering research', *Proc. I.M.E.*, vol. 163, 1950, p. 27.

Browne, R. C., *Health in Industry*. Edward Arnold, London, 1961. '

Clenshaw, W. J., 'Measurement of strain in components of complicated form by brittle lacquer coatings', *Proc. I.M.E.*, vol. 152, 1945, p. 221.

Constant, H., 'Early history of the axial type of gas turbine engine', *Proc. I.M.E.*, vol. 153, 1945, p. 411.

Galloway, D. F., 'Machine tool research, design and utilization', *Proc. I.M.E.*, vol. 175, 1961, p. 85.

Gardner, G. W. H., 'Guided Missiles', *Proc. I.M.E.*, vol. 169, 1955, p. 30.

Gartmann, H., *Science as History*. Hodder and Stoughton, London, 1960.

Gibb, Sir C. D., 'Some engineering problems in connection with the industrial application of nuclear energy', *Proc. I.M.E.*, vol. 171, 1957, p. 22.

Gibbs-Smith, C. H., *The Aeroplane*. Sc. Mus. Pubn., H.M. Stationery Office, London, 1960.

Giedion, S., *Mechanization takes command*, 2nd edn. O.U.P., 1955.

Farmer, H. O., 'Free piston compressor engines', *Proc. I.M.E.*, vol. 156, 1947, p. 253.

Fessler, H., and Rose, R. T., 'Photo-elastic investigation of stresses in the heads of thick pressure vessels', *Proc. I.M.E.*, vol. 171, 1957, p. 633.

Hawthorne, W. R., 'The early development of the Dracone flexible barge', *Proc. I.M.E.*, vol. 175, 1961, p. 52.

Kohler, J. W., and Jonkers, C. O., 'Fundamentals of the gas refrigerating machine', *Philips Technical Review*, vol. 16, 1954, p. 69.

Larsen, E., *Atomic Energy*. Pan Books Ltd., London, 1958.

Macmillan, R. H., *Automation, Friend or Foe?* C.U.P., 1956.

Newman, R. P., and Houldcroft, P. T., 'Welding—engineering and metallurgical aspects', *C.M.E.*, vol. 8, 1961, p. 214.

Sikorsky, I. I., 'The transport helicopter', *Proc. I.M.E.*, vol. 169, 1955, p. 1183.

Whittle, F., 'Early history of the Whittle jet propulsion gas turbine', *Proc. I.M.E.*, vol. 152, 1945, p. 419.

Wrangham, D. A., *The Elements of Heat Flow*. Chatto and Windus, London, 1961.

Index

Index

454